Desirable Residences
and Other Stories

Desirable Residences

and Other Stories

E. F. BENSON

Selected by
Jack Adrian

Oxford   New York
OXFORD UNIVERSITY PRESS

Oxford University Press, Walton Street, Oxford OX2 6DP

Oxford New York Toronto
Delhi Bombay Calcutta Madras Karachi
Petaling Jaya Singapore Hong Kong Tokyo
Nairobi Dar es Salaam Cape Town
Melbourne Auckland
and associated companies in
Berlin Ibadan

Oxford is a trade mark of Oxford University Press

First published 1991
Reprinted 1991

British Library Cataloguing in Publication Data
Data available

Library of Congress Cataloging in Publication Data
Benson, E. F. (Edward Frederic), 1867–1940.
Desirable residences and other stories / E. F. Benson; selected by
Jack Adrian.
p. cm.
I. Adrian, Jack. II. Title.
PR6003.E66D4 1991 823'.912—dc20 91–9923
ISBN 0–19–212304–1

Printed in Great Britain by
Biddles Ltd.,
Guildford and King's Lynn

Acknowledgements

MOST of the stories in this collection have been discovered while trawling through fiction magazine volumes in that most civilized of repositories, the Bodleian Library, Oxford. I should like to acknowledge my debt to the staff in the Old Library Reading Rooms, the Copying Department, and the Stack—in particular to R. J. Roberts, Deputy Librarian and Keeper of Printed Books.

I should also like to acknowledge the aid of the following: Mike Ashley, who pointed me in the direction of the *Lady's Realm* and thus to the two lost 'Dodo' stories; my old friend Bill Lofts, who often undertook arduous journeys to the Newspaper Library at Colindale on my behalf; George Locke, whose acquisition at auction of 'Number 12' and 'The Top Landing' in typescript spurred me on to discover them for myself in the pages of *Eve*; Cynthia Reavell of the Tilling Society, who very generously provided the copy of 'Desirable Residences' (originally discovered by Malcolm Baker) and much useful information and advice; and, finally, Michael Cox, who has had to put up with much.

There are currently two societies devoted to the life and works of E. F. Benson: the Tilling Society (Organizers: Cynthia and Tony Reavell), the Martello Bookshop, 26 High Street, Rye, East Sussex TN31 7JJ; and the E. F. Benson Society (Secretary: Allan Downend), 88 Tollington Park, London N4 3RA.

Contents

Introduction

FOR three or four decades after his death in 1940, E. F. Benson was one of the forgotten men of English letters, only remembered (if at all) by supernatural fiction enthusiasts, who paid high prices for copies of his four elusive ghost-story collections: *The Room in the Tower* (1912), *Visible and Invisible* (1923), *Spook Stories* (1928), and *More Spook Stories* (1934). In these circles Benson's reputation remained high, his skill at being able (in his words) to 'induce in the reader an uneasy frame of mind which . . . paves the way for terror' justly celebrated. Inevitably, he was bracketed with that other great practitioner of the classic English ghost story, his friend M. R. James, although Benson's stories had a much harder edge, and in the matter of describing supernatural horrors he was rather less squeamish than his older contemporary.

By contrast, Benson's other writings—in particular his waspish comic novels about meddlesome female social climbers, such as his 'Mapp and Lucia' cycle—were ignored by all but a tiny, though vocal, core of readers—including Noel Coward (an enthusiastic Bensonite) and the American critic Gilbert Seldes. Over the past decade all this has changed. Two successful British television series based on the Mapp and Lucia novels, a number of splendid readings on BBC radio by the actor Aubrey Woods, and a flood of reprints of his best comic novels have transformed E. F. Benson into a popular writer once more, with not one but two societies devoted to his memory.

Accomplished though he was as a comic novelist, Benson was also at home with the demands of the short story. During the last twenty years of his life he published only collections of ghost stories; indeed, he became so well known as a writer of what he called 'spook' stories that he seems to have given no thought to gathering together the large number of other short tales he wrote during a long career as a professional writer. Benson never lived less than comfortably by his pen, taking great care to tailor his work to the requirements of specific periodicals, publishers, and editors. He was particularly prudent in maintaining a good working relationship with the latter, especially magazine editors. Most of his longer fiction appeared first in serial form, and when his novels were issued as hardbacks it was invariably by publishers (Heinemann, Cassell, Hutchinson, Hodder & Stoughton) who supplied novels to the superior end of the circulating and subscription library markets, which bought fiction in bulk—vital for the continuing well-being of both publishers and authors. Similarly, Benson's short stories appeared mainly in the grander periodicals: glossy Society weeklies (the *Illustrated London News*, *The Tatler*, *Eve*); reputable monthlies (*Pearson's Magazine*, the *Windsor Magazine*, the *Pall Mall Magazine*, the *Cornhill*); and the more dignified women's magazine's (*Lady's Realm*, *Woman*, *Women at Home*, *Lady's Field*).

Benson sold much of his work through an agent, but this did not prevent him building up friendly relations with two magazine editors—unrelated but both called Hutchinson—that were to prove crucial during the 1920s when Benson was shifting from creaking melodrama to comedy.

Walter Hutchinson was a man of genuinely inspired ideas, which, however, often went some way beyond what his innate parsimony would allow. Hutchinson, his firm, specialized in popular middle-brow library fiction, as well as producing a small though lucrative line in part-works. Before paper came off the wartime ration list in 1919, Walter astutely foresaw the subsequent boom in periodical publishing and launched *Hutchinson's Magazine*, which within five years became the flagship of a flourishing magazine empire aimed at all brows—from high (*Light: A Journal of Spiritual Progress* and *The Smart Set*), through middle (*Sovereign Magazine* and *Woman*), to the determinedly low (*Action*, *Mystery Story* and *Adventure Story*, all of which carried a preponderance of second rights material from American pulp magazines).

Most of Walter Hutchinson's magazines were heavily dependent on fiction; but in 1923 he changed course, transforming *Hutchinson's* into an enlarged 'fine art' production printed on heavy stock. The best popular authors and journalists wrote for him, the best commercial artists provided illustrations. In this Hutchinson again showed his shrewdness by anticipating—and influencing—such later best-selling general interest magazines as *Nash's Pall Mall, Woman's Journal, Britannia* and *Eve*. Unfortunately, Hutchinson's incorrigible penny-pinching sabotaged the enterprise and by the end of the decade all that remained was *Hutchinson's* itself, no longer printed on art stock, drastically reduced in dimensions, and visibly moribund. Few first-rank writers now contributed to it, and what illustrations there were appeared to be the work of an office junior.

In what seems to have been the last issue of *Hutchinson's* (for December 1929) a story by Benson appeared: in the circumstances, striking evidence of the author's loyalty to the man who had issued a number of his best comic novels (including *Miss Mapp, Lucia in London*, and *Paying Guests*), as well as what are arguably some of the finest ghost stories of the twentieth century. In the early 1920s Benson had enjoyed several markets for his supernatural fiction, but in 1922 Hutchinson, who had a weakness for such stories, had persuaded him to direct his energies towards *Hutchinson's*, with the result that for the next seven years Benson produced nearly thirty 'spook' stories, many of them classics of the genre.

The second Hutchinson, Arthur ('that wise . . . good man', as Dornford Yates—not noted for charitable assessments, especially of editors—once remarked), presided over the *Windsor Magazine*, which was aimed directly at the leisured upper middle class and specialized in superior, highly polished romance. There was nothing in the *Windsor* that was controversial or disagreeable, nothing experimental. It was the perfect companion for a train journey from Paddington to Penzance—ideal reading matter for an archdeacon's wife. Benson first came into contact with its publishers, Ward, Lock, when they bought up what remained of the wreck of A. D. Innes, the firm that had published Benson's novel *Limitations* in 1896 (as well as works by Anthony Hope, Eden Phillpotts, J. C. Snaith, Basil Thomson, Stanley J. Weyman, Francis Gribble, and other coming men of the 1890s). Benson found the *Windsor* (for most writers a notoriously difficult magazine to break into) an ideal market for his light, slightly acerbic tales of bustling social intrigue in spas, small country towns, and amongst the metropolitan cultural set.

To Arthur Hutchinson, Benson was a valued contributor. Over half the stories he wrote during his thirty-year association with the *Windsor* appeared in the coveted December (i.e. Christmas) or January issues. During the 1920s Benson was especially prolific, and it seems likely he would happily have continued dashing off three or four tales a year for Arthur Hutchinson until the day he died. Unhappily, it was Hutchinson who died, in harness, in 1927. Benson's final original contribution to the *Windsor* appeared in the January 1928 issue—a story almost certainly written before Hutchinson's death since his next in the *Windsor* vein, 'The Guardian Angel' (included here), appeared in the rival *Woman* (which had previously featured some of his ghost stories) in April. Clearly, the new editor of the *Windsor* was unsympathetic to Benson's peculiar talents, and Benson himself turned to other periodicals—of which, with the collapse of Walter Hutchinson's empire and the crash of the popular short-story market in general, there were dwindling numbers. As a result, in the last ten years of his life Benson's output of short fiction was sparse.

Benson was never considered 'good enough' for the *Strand Magazine*, most prestigious of the middle-class, middle-brow periodicals, which regularly featured the effusions of such seasoned romantic melodramatists as Ethel M. Dell, Gilbert Frankau, and Warwick Deeping, and in any case favoured the 'silly ass' style of comedy supremely sired by P. G. Wodehouse. A single, and indifferent, story by Benson was accepted in 1920—significantly, at a time when the *Strand* was still casting around for contributors to replace those who had died in the First World War. Greenhough Smith, the *Strand*'s editor, demanded concrete characterization, plenty of 'colour', a strong sense of place, and, above all, plot—preferably with a sting in the tail. Benson's comic characters were always skilfully contrived, and his local colour—especially if a story was set in places such as Greece or Egypt that he knew well—was often impressively deployed; but plotting was not his best suit. His sharpest stings were dabbed into the fabric of a story rather than sprung all at once at the end, and his denouements were frequently telegraphed long before the final chapter or paragraph.

What Benson excelled at was the *feuilleton*, a type of serial much favoured in the Edwardian period, in which the main plotline does not curve gracefully from start to finish, over a succession of integrated sub-plots that are set up and then resolved as the story progresses. Instead, the story is told episodically, with each segment packed with interest and incident but not necessarily following through to the next instalment, or indeed to the finale. This method came to full flower in Benson's comic novels of the 1920s and 1930s, principally in the Mapp and Lucia books, which are virtually plotless, consisting instead of a succession of farcical incidents, preposterous catastrophes, and small wars in which first Mapp wins, then Lucia, then Mapp, then Lucia, until a satisfactory state of armed truce is finally achieved, with Lucia just gaining the upper hand.

In 1926 the journalist Beverley Nichols (who had taken upon himself the role of Young God toppling the Old, of whatever persuasion) rather meanly insinuated in the pages of the Society weekly *The Sketch* that Benson was past it, that his works were 'gathering dust' on library shelves. Nichols's tone was offensively trilling and the judgement unkind; but there was nevertheless an element of truth in it. Most of Benson's novels—the Mapp and Lucia cycle apart—are Edwardian melodramas rather than stories of the Jazz Age, and his most celebrated book, *Dodo* (1893), which recounted the escapades of a relentlessly witty but essentially heartless Society girl, was a comedy firmly of the late Victorian era. And although Benson had always been known for his humour, in many of his novels there was often an uneasy juxtaposition of the hilarious and the mawkish.

Queen Lucia (1920) and *Miss Mapp* (1922) signalled a change of direction—no religiosity, no embarrassing deathbed scenes, no too-perfect heroes and heroines. The change was recognized by most critics, although Katharine Mansfield, reviewing anonymously in *The Athenaeum*, dismissed *Queen Lucia* with the comment that Benson's humour had gone 'not to the dogs—but to the cats'. But Mansfield was a lone dissenter, and from the mid-1920s Benson concentrated with increasing success on this new vein of pure comedy, as well as his tales and novels of the supernatural. It was almost as though, after nearly thirty years as an author, he had suddenly found his true style.

Like all professional writers, Benson seldom wasted a good idea and on occasion expanded short stories into full-length novels: 'To Account Rendered' (included here) clearly sparked off the theme and climax of *Travail of Gold* (1933); his 1929 horror story 'The Wishing Well' (from *More Spook Stories*) contains the germ of what is surely his most bizarre novel, *Raven's Brood* (1934). Similarly, many of the incidents and ideas in the lightly astringent confections he wrote for Arthur Hutchinson—particularly the Amy Bondham stories, some of which are collected here—were later recycled in the Mapp and Lucia saga.

Of the twenty-seven stories in this collection only two have appeared in book form before: 'Philip's Safety Razor' (which lightly mocks a husband and wife team of *feuilletonists*, almost certainly based on Heath Hosken and Coralie Stanton, prodigiously prolific suppliers of serials for the *Daily Mail* before and during the First World War) in Benson's *The Countess of Lowndes Square* (1920) and 'Aunts and Pianos' in Lady Cynthia Asquith's anthology *The Funny Bone* (1928). The rest have never been reprinted and exhibit all aspects of Benson's talent, from the glittering but unfeeling prattle of Dodo (in two rare stories from the 1890s) to the delightfully venomous point-scoring in which his more calculating female characters are wont to indulge; from freaks, cranks, and obsessives to terrors that lurk both by day and by night.

The title story would appear to be the only short story in the Mapp and Lucia cycle to have escaped collection in book form. A minor editorial adjustment has been made, for neatness' sake. The story takes place in Tilling, which is of course Rye, where Benson lived more or less permanently from 1920. He frequently used the town in his novels and stories, renaming it variously Tarleton, Trenthorpe, Trench, Scarling, and so on. Sometimes he called it Tillingham (as in the Amy Bondham story 'Entomology', reprinted here, and in 'The Witch-ball', reprinted in *The Flint Knife: Further Spook Stories*, 1988). Generally, however, the name Tilling was reserved for stories concerning Mapp and Lucia; yet when 'Desirable Residences' first appeared in *Good Housekeeping* in 1929 Miss Mapp and her dumpy sparring-partner Diva Plaistow are suddenly residents not of Tilling but of Tillingham. I have had little hesitation in excising the 'hams' and silently removing Miss Mapp and Diva to what all readers of E. F. Benson must feel to be their proper location.

JACK ADRIAN
February 1991

THE DIVERSIONS OF
AMY BONDHAM

ENTOMOLOGY

THE more intelligent sections of London society were beginning to get a little tired of the discussion of plays which dealt exclusively with drunkards, drabs, drug-fiends and other disreputable folk. Of course—so the intelligent declared—they were marvellously like life and so human, but after seeing a certain number of them you got satiated with these particular sides of life and humanity. Vice taken in undiluted tumblerfuls palled on the jaded palate.

Amy Bondham, whose career was to collect round her all who were famous for intelligence, artistic ability, social distinction, or even birth, had never really cared for these plays. Naturally she had to go to all the first nights, and run round to green-rooms afterwards to congratulate actors and authors, and secure their presence at lunch or supper in her house, but it was a relief to her strong moral sense of uprightness, sobriety, and the domestic virtues, when this slight waning of interest in the affairs of menageries and monkey-houses began to manifest itself. She had been quick to perceive it, and with equal quickness she realized that artistic circles in Chelsea, earnest circles in Bloomsbury, and smart circles in Mayfair would presently be looking about for something fresh. She must therefore exercise her utmost powers of intuition to divine what the new angling would be, and bait her hospitable hooks for the capture of the largest fish.

It was about a month before Easter when she consciously registered these impressions, and she kept her ears alert to catch the first faint tokens that should herald the advent of the forthcoming fad. There were whispers of a tremendous pianist who was to deafen London during June, of a troupe of marvellous artists belonging to a

West African cannibal tribe, who were to present some very stimulating plays, of an Italian singer who could sustain with resonance and clarity a shake on the top two notes of the piano, of an operation which would not only restore but perpetuate youth, but in none of these entrancing events did she seem to feel any assured confidence. There had been so many pianists, so many singers, so many operations, and though a play by West African cannibals was a novelty, there would be risks in asking these artists to her house. They might be very hungry or very homesick, and who could tell what might happen then? She would never forgive herself if any of her guests were killed and eaten.

She was stopping at home one afternoon, nursing a cold that was the result of having attended a distinguished marriage, immediately followed by a distinguished funeral in a snowstorm. But a day in the house was never wasted, and a dozen notes to various prominent people had brought her a succession of desirable visitors. The first of these had said something about the New Psychology, and she had not particularly attended, but when an hour afterwards a member of the late Labour Government alluded to the same subject, she began to listen and asked intelligent questions. It sounded rather fascinating, and Mr Crabbit had the gift of lucid statement.

'It is frankly materialistic,' he said, 'and the general theory is that every mental gift is a sort of emanation from a physical organ. A man of altruistic nature will literally have fine action of the heart, courage depends on the spleen, optimism on the liver, and so forth. So instead of cultivating your mental gifts, you cultivate the physical organs on which your mental gifts depend; you take strychnine to cure you of selfishness, and a pill for pessimism.'

She faintly heard the clarions heralding an august approach; this was just the sort of thing which would interest London.

'How marvellous!' she said. 'And who is the author or discoverer of "the New Psychology"?'

'A professor from one of our minor universities,' he said. 'What is his name now? . . . Ah, yes, Vincent Fleet. There is some talk of his giving lectures in London on it during May.'

The succession of her visitors went on: she had asked them at intervals of half an hour, so that there would be a little pleasant overlapping and yet a little tête-à-tête with each. They all seemed to have heard of the New Psychology; the New Psychology was clearly in the air. And then with the arrival of Miss Elsie Dane, the distinguished cinema star, there came an exceeding great reward. It was hardly a stroke of luck; it was the reward that always attends the alert and inquisitive. Miss Elsie Dane was voluble.

'I'm run down, that's what I am,' she said, 'and when last week I was down at Tillingham, featuring Cleopatra, I determined to take a month off and spend it there. You never saw such a place—a little red-brick town in the middle of flat marshes, with a slow river winding out to sea. The river was the Nile, and Cleopatra and Mark Antony were fishing. But that's the place to have a rest in. It got left behind two hundred years ago, and has remained there ever since, exactly as it was. There was a house there that I fell in love with, too. You never saw such an abode of peace, right up at the top of a cobbled street where nobody ever comes, with a walled-in garden behind, and I just saw myself dozing and vegetating there for a month. And such a name, too—the name of it was Slepe House, though you'd hardly believe such a thing was possible.'

'How delicious!' said Amy, who was getting rather tired of this. 'And are you going to take it?'

Next moment she attended violently again.

'Why, no, there was the ill-luck of it,' said Elsie. 'It had been to let, but the very day before I got there it had been taken for the month of April, just when I wanted it, by some frowsy old professor called Vincent Fleet. Too bad, wasn't it? The house-agent told me that he was coming down for a month's complete quiet to prepare some lectures he is giving in Town later on.'

'Dear me, I seem to have heard of him,' said Amy. 'He's the exponent of the New Psychology, I believe.'

'That's it. A pack of rubbish, I expect. And I shall have to look out for another house. You don't know of one in some backwater, do you?'

During the next month the dawning interest in the New Psychology began to brighten in the most wonderful manner. Amy got a small manual about it by Vincent Fleet, and, having read it, procured a dozen more copies, which she distributed to friends of light and leading. The movement was catching on; everybody was talking of it, and nobody talked more than Amy Bondham. With characteristic prudence she said nothing whatever about Slepe House or her plans for Easter, but she thought of them thoroughly in bulk and detail, and engaged a single room at the famous old inn at Tillingham. She also bought a small water-colour paint-box and a sketching-block.

She and her pink, amorous little husband, who was still convinced—and who shall say that he was wrong?—that his Amy was the most wonderful woman in the world, had arranged to spend an Easter fortnight at Le Touquet, where with great content and perspiration he dug up quantities of small pieces of turf from the golf links. They were dining—by exception—alone one night, and she sat herself on the hearthrug and leaned against his knees.

'I've got a terrible confession to make, Christophero mio,' she said, 'and I am so ashamed of it. But I do feel very slack and tired, and I don't really feel up to coming to Le Touquet and having all those delicious games of golf with you. I want to go away to some quiet little place where there will be nothing to do and nobody to see.'

'That'll suit me,' said he. 'Better telegraph and give up our rooms at Le Touquet. Where shall we go, pet?'

'No, I shouldn't dream of allowing you to give up your fortnight at Le Touquet,' she said. 'I know how you enjoy it. And you would be so bored with the sort of place I'm thinking of. You must promise me you'll go to Le Touquet, and write to me every day, or else I shall come, too.'

'I would much sooner come with my Pussikins wherever she goes,' said he.

'But I don't wish you to. I want to be utterly idle and indolent. I want not to talk, and if you were there I should go on chattering eternally to you, as I always do. I want to be a vegetable.'

'And where are you thinking of going?' asked he.

'Well, I thought possibly of going to Tillingham. Elsie—dear Elsie Dane, you know—told me it was an ideally restful place. I shall just take a room at the inn, and get up late and sit about, perhaps sketch a little.'

'But do you know anyone there?' he asked.

'No, darling. That's why I want to go there.'

This was all very unlike his Amy, who usually spent the holidays wherever she might expect the greatest and smartest crowd of people; but, like the adoring man he was, he went off obediently to Le Touquet, and she the same afternoon went down to Tillingham. The antique little town stood glowing red in the sunset, and she instantly set out on a swift exploration before dark. It was all tranquil and delicious and old-world, and most sequestered of all was the cobbled street up which, in accordance with information received, she was presently walking. At the corner of it, where it turned sharply towards the church that crowned the hill, was the house she was looking for, red-fronted and Queen Anne in date, and there on the brass plate on the door was the gratifying intelligence that this was indeed the Mecca of her pilgrimage. She went back to the High Street, where a stationer's shop was still open, and bought a picture postcard of Slepe House. It was easier to copy this than to sketch freehand from the house itself, and a couple of hours' work after dinner resulted in a fairly successful enlarged reproduction of it on her sketching-block.

Now, in the weeks that had intervened between the revelation of Elsie and Amy's arrival at Tillingham she had gathered a few facts about Vincent Fleet, to handle which required the utmost ingenuity. He was middle-aged, unmarried, of an intensely shy disposition and of very recluse habits. She could only find one person who (a cousin of his) knew him, and this lady frankly told her that it was no imaginable use to give her a note of introduction to him, because he would not take the slightest notice of it. It had been tried before, and nothing but disappointment had followed. Now, too, Mr Fleet was in strictest hermitage over the lectures he was preparing, and more than ever would he practise this isolation which lost him so many pleasures, of which the acquaintance with Mrs Bondham was assuredly not the least. And this unobliging cousin, whom Amy had once rather snubbed before her relationship had been of any value, was unkind enough to write a little letter to Mr Fleet, warning him of the possible advent of the nimblest climber in all London. This Amy did not know, but it would have made very little difference to her if she had.

Amy had resolved to begin quietly. She was perfectly capable of beginning violently, if time was short, and if she thought that a position could be rushed. She had made several notable captures that way at bazaars opened by Royal personages. But on this occasion there was plenty of time. She had a month to spend at Tillingham, and so had Vincent Fleet. She had formed the impression that beleaguering methods were likely to pay better than direct assault, and consequently next morning she encamped on the pavement opposite Slepe House with a canvas-seated stool and her drawing materials. Her sketch of the lower storey of the house, with its eaved and massive door, was already faintly outlined, and it was pleasant to sit in the mellow April sunshine and cast glances at its windows. Perhaps nothing would happen that morning to give a clue to the habits of its tenant, and with her admirable thoroughness she was prepared to come back in the afternoon and sit there again. He would surely come out some time during the day.

It was half-past twelve when the front door opened and a big, lean man emerged. He had a plain and pensive face, gave her one glance and walked rapidly down the narrow street with soft padding footsteps: he had evidently rubber soles to his shoes. Half an hour later, precisely as some clock struck one, she heard soft padding footsteps coming from the direction of the church behind her, where she sat at this sharp angle of the street, and he flitted by, passing close to her, and entered his front door. Next day exactly the same thing happened, and, as she heard the footsteps of his return, she held out her sketch, already dabbed with colour, at arm's length, as if to realize the effect of her work. She hoped that he might pause, as he went by, with the natural curiosity of a man who sees a picture being made of the house he inhabits. She thought he did, and looked up with a very winning smile. The pavement where she sat was narrow, and she accompanied her smile by saying: 'I'm so sorry: I'm afraid I'm obstructing the pavement.'

He raised his hat, gave an indistinguishable murmur, and seemed to be in a hurry to regain the shelter of his house. That was like what she had heard of him and his shyness which deprived him of so many pleasures. But two days' devotion to art had given her only the meagre result that Vincent Fleet went out at half-past twelve and returned at one. She could also conjecture that he lunched very soon afterwards, for the sound of a gong in the house a few minutes later probably indicated food.

But she was still in no hurry, and it was delightful to stray about these picturesque streets and out into the broad marshlands beyond the town. She made a study of the walls of the spacious garden belonging to Slepe House, and discovered that a door led from it into a quaint narrow alley that debouched into the street opposite her inn. Quaint, too, were the names of the shops and offices there. On one side of this alley was a photographer called Frostbite, on

the other the offices of Crimp and Pantile, solicitors. Once, as she
returned from a country ramble, she thought she saw Vincent Fleet
coming out of Crimp and Pantile's office and turning sharply up the
alley which led to the garden door, but the light was bad, and she
could not be certain.

A fruitless week passed, and Amy resolved on more decisive
methods. She was still sketching in front of Slepe House, from a
slightly different angle, and this morning, just as he returned on the
stroke of one, she dexterously upset her camp stool and fell on the
pavement with her leg under her. It hurt more than she expected,
but there was instant balm.

He hurried to her. 'Dear me, I'm afraid you have had an awkward
fall,' he said.

Amy gave a writhing smile. 'My ankle,' she murmured, and she
attempted to rise. 'It's nothing.'

He became quite peremptory. 'Indeed, you mustn't think of trying
to stand on it,' he said. 'You must please allow me to get one of my
servants, and we'll help you into my house. We'll move you quite
easily without your putting your foot to the ground. Most important
not to put any weight on it.'

'You're too kind,' she said. 'But I'm staying at the Royal, not a
hundred yards away, and I'm sure I could hobble there.'

'I must beg you not to attempt anything of the kind,' said he.

The pain was rapidly passing off, but Amy writhed again, and was
very brave when he and a parlour-maid made a chair of crossed
hands for her and carried her into her Mecca. She was placed on a
sofa in a delightful little dining-room, looking on to the garden,
where lunch was already laid. The parlour-maid unlaced her shoe
and took it off, and she winced beautifully and turned her grimace
into a smile.

He was looking anxiously at her, and, for all his shyness and
recluse habits, she felt sure he had a tender heart. There was the
brow of the thinker, the wide intuitive eyes, the sensitive mouth.

'Now I'm going to telephone for the doctor at once,' he said, 'and
you mustn't move till he has seen you and examined your poor ankle.
I insist.'

While he was gone she looked eagerly round the room. One table
was laid for lunch, another in the window was strewn with books and
manuscripts. She burned to glance at them, but it was unsafe to

leave the sofa, for the psychologist might return at any moment. He came back with the joyful news that Dr Blaistowe was out.

'But really I must keep you a prisoner,' he said, 'till he has seen you. A quite small injury may give a lot of trouble if you take liberties with it. I hope, if you feel up to it, you will have your lunch with me.'

She felt up to it, and her couch was wheeled to the table. Remembering his shyness, she was far too wise to allude to the New Psychology, or, indeed, to hint that she knew who he was, and he, with the childlike simplicity of the great, did not introduce himself. She was pleased at that: he would assume that she thought he was a mere local little householder, and no idea that she was playing up to his distinction would enter his head.

'Your delicious Tillingham,' she said, 'how happy I have been here all alone, just resting and sketching and drinking in the sweet tranquillity of the place! I was terribly tired when I came down from London, but I've been bathing in its peace. I wouldn't even allow my husband to come with me. I sent him off, in spite of his protests, to Le Touquet.'

That was a wise thought; it was well to strike the keynote of respectability firmly at the outset. She proceeded:

'London has been such a whirl ever since Christmas,' she said, 'and I felt I must get away alone and cool down, so to speak. And what a wonderful place Tillingham would be for anyone who wanted to do some great piece of work!'

That conveyed a subtle invitation to him to disclose himself if he wanted. Apparently he didn't want.

'That is quite true,' he said. 'I have had a good deal on my hands lately, and it is wonderful how much one can get through with no external distractions.'

This was getting a little closer, but she felt sure she was wise not to press him. It was time, however, now to acquaint him with the brilliance of her own existence, and she launched out on to the dazzling sea of the names of those among whom she moved. There were scientists and artists there, Prime Ministers and playwrights, alluded to mostly by their Christian names, and the tie that bound them together and her to them was that they were all Thinkers. There were naturalists, too, students of bird-life and botany and butterflies. . . . He was looking a little dazed, but at the mention of butterflies he recovered himself.

'I have studied entomology all my life,' he said. 'I wish I could have devoted my whole time to it. But there is my work. That has to come first; these wonderful insects are only my hobby.'

Amy evinced the utmost interest in moths, and he was in the middle of telling her curious things about the colours on their wings being often not pigments at all, but prismatic effects of light on innumerable minute striations, when the doctor was announced. Her host delicately withdrew while her ankle was examined, and returned to hear the most encouraging report. The twist she had given it was a very slight one; she might, without fear of evil consequences, walk back to her hotel.

'I wish it was more serious,' she magnificently exclaimed, when Dr Blaistowe had gone. 'I could lie here and listen to your wonderful stories all day.'

The time had come.

'Now you must allow me to follow up this chance that has brought us together,' she said. 'I shall be back in London at the end of this month, and I shall be so honoured and proud if you will come and dine with me any night after that, for I shan't dream of intruding myself on you here again. I know how busy you are, and in my small way I realize how necessary solitude is for concentration.'

She found a card and scribbled her telephone number in the corner.

'I'm Mrs Bondham,' she said, 'though how unlikely that you should ever have heard of me! But it would be such a pleasure to get a few friends to meet you any night you choose.'

He looked slightly surprised. 'But that's most kind of you,' he said. 'I shall be going up to London next week, and I shall stop there until the middle of May. In fact, I've let my house till then—a Mr Vincent Fleet, who, like you, is coming here for quiet. I fancy he is giving some lectures in London later on, for which he is preparing.'

There was a moment's silence, and her host took a card out of a case on his littered table.

'He was coming down here a week ago,' he said, 'but postponed his visit. . . . May I have the honour of presenting you with my card?'

Amy glanced at it. She read:

MR WILLIAM PANTILE
Slepe House,
Tillingham.

She required no more than a couple of seconds to pull herself together.

'You've been too kind to me, Mr Pantile,' she said. 'Mind you don't forget to ring me up when you come to London. You must come and have tea with me, and tell me more about moths.'

But Amy's holiday was not quite thrown away. Vincent Fleet gave his series of lectures on the New Psychology, and they proved the rage of the season; nobody talked about anything except him and his amazing philosophy. Amy did not actually succeed in meeting him, but her conversation was full of him.

'A marvellous mind,' she said. 'And you've no idea how delightful he is alone. Dear Vincent! He took a charming little house down at Tillingham, where I spent Easter—one of those exquisite Queen Anne houses, with a little square dining-room looking on to a garden with a high wall round it. Slepe House. A table in the window, where he worked, covered with books and manuscripts. I never enjoyed such a tranquil and illuminating time.'

THE PEERAGE CURE

IT had been the most wonderful autumn for Mrs Amy Bondham: never before had she lived so exclusively in the society of the great and the ennobled. She had spent October in a round (you might call it a merry-go-round) of visits; for though she had originally planned only four weekends, she had made herself so popular at each of them that some member of the party had insisted, or at least consented, that she should spend the intervening days before her next engagement at a Castle or a Grange or something very moated and hereditary. Then, when the world began to stream back to London again in November, it was her turn, and every day the hospitable table in her house in Mount Street was laid for intimate and high-born little parties. Though still fond of professional distinction, she had dropped, but only temporarily, her literary, artistic, and histrionic friends, whom she meant to pick up again in the close time for the high-born circle which flocked to the Riviera after Christmas.

Just now her aim and ambition was the high-born, and the pages of her 'Peerage', in which she put a little cross in the margin opposite the names of those who had entertained her or been entertained by her, became full of these discreet little pencil-marks, and pleasantly swollen became the album of picture postcards into which she gummed, with absolutely scrupulous honesty, only the photographs of mansions which she had actually visited, inscribed with the appropriate date. The rumour, therefore, that she bought these in indiscriminate packets, as illustrating the houses at which she would like to stay, was an unfounded fabrication, devised by the jealousy of less fortunate competitors.

Her circular and devoted little husband Christopher had accompanied her on her progress in October, but he was not so strong as she, and also was quite unable to resist the pleasures of the table. Throughout November, in spite of the obedient walks he took daily round and round the park, he remained very liverish and gouty; and when, early in December, Amy was about to set forth on the visit which really was the crown of all her social attainments, he was persuaded by her to go off to Bath, and try to get fit again for the Christmas Campaign. This crowning visit was to the Duke of Whitby, and though it was certainly now coming off, it had required all Amy's tact and perseverance to effect it, for the Duchess had continued to be impervious to her hints for an unusually long time. But finally doggedness carried the day, the Duchess had yielded, and had asked her to Doncaster Castle while she was entertaining a bevy of distinguished savants from the Psychological Conference in York. It was not precisely the sort of party which Amy would have chosen, but if it had been a party of chimney-sweeps and chiropodists she would have eagerly accepted. She could murmur her 'Nunc Dimittis' now, and see about all the clever people she had dropped.

She and Christopher dined alone for the first time since September, as Amy delightedly remembered, on the evening before he went to Bath and she to Doncaster. For once she had broken her rule about the picture-postcard album, and had allowed herself to buy four striking views of the magnificent Norman pile in which she would be dining next night, and subsequently sleeping, if excitement would allow her to do so.

'I may as well put them in now,' she said, 'because they will just fill up the last page in my book. I must get a new one when I come back, for we're going to Eagles for Christmas and Tenterden Grange for the New Year.'

'Better not, Posie,' he said. 'It may bring bad luck. The Duchess may put you off tomorrow morning. You've not got there yet.'

She laughed.

'You superstitious old man!' she said. 'That's gout. It's just acidity. Morbid ideas like that are purely physical. Where's the gum-pot? My dear, what a wonderful autumn it has been. Look, the book was nearly new in September.'

She turned over the rich pages.

'All those!' she said. 'I think I shall have it bound in a manner more worthy of its contents. I wonder what we shall have in the next volume?'

She came and sat on the hearth-rug, propped her back against the arm of his chair, and stretched her feet out to the blaze.

'Perhaps there won't be a next volume,' she said, 'for really, Christopher, I feel I've been very frivolous all the autumn.'

The words 'perhaps there won't be a next' gave Christopher a queer little shudder, but that probably was acidity too.

'You bet there will, Posie,' he said, 'if there are enough fine houses left in England to fill it. You have become a fashionable little dame.'

She sighed.

'But there are other things besides that,' she said, rather doubtfully. 'There's that volume of Proust which I must read, and *The Life and Times of Tutankhamen*, and that book on Auto-suggestion. I shall take them up to Yorkshire. And when my visit's over I shall join you at Bath.'

'Better not do that,' he said. 'You'll be bored to death.'

She considered.

'Well, we'll see,' she said. 'I've noticed sometimes in the paper that there are interesting people at Bath—and some interesting houses in the neighbourhood,' she added.

There had been a week's frost in England, and Amy, next morning, seeing in the paper that there was skating in the north, decided to take her skates with her. She was quite an expert on the ice, having spent the last winter at St Moritz, where she had come across a great many agreeable people, and had, in fact, laid the foundation of the superb autumn she had just enjoyed. One of her picture postcards also showed a lake below the walls of the Castle, and another a medieval moat round it. Probably the lake or the moat would bear, and the idea of discussing the newest views on Auto-suggestion with eminent psychologists, and then breaking off to astonish them by her lissome feats on the ice, was very attractive. Like most of her plans this turned out well, for the Duke was an ardent skater himself, and after opening the Psychological Conference with a weighty speech, he refused to attend any more meetings, and stopped at home in order to waltz with Amy on the ice-covered moat. His secretary was an adequate pianist, and he was bidden to neglect all his business and play dance music for them. Wrapped in a fur coat, this unfortunate young man sat by the open window of the pink drawing-room so that the lively strains might reach the dancers, while Amy and the Duke pirouetted all day on the frozen water of the moat immediately below. One evening a Royal Princess dined at the Castle, and Amy grew greater than ever, for they skated again after dinner, and she nestled against the Riband of the Garter. She sat up half the night writing one account of all this to poor Christopher at Bath and another to a struggling young friend of hers who wrote paragraphs for the Press. She would make half a dozen paragraphs out of such material, and Amy, though yawning her head off, did not go to bed till she had fully completed this act of disinterested kindness.

The day of her departure, already twice postponed, arrived, and the pain of parting was slightly lessened by the fact that a thaw had set in. The Duke, however, said that the ice would hold for the morning, and they swished about in ever-deepening puddles of water. Ominous crackings and bubblings of air at last warned them that the ice was safe no longer, but then it was too late. A piece collapsed, and they were left standing in thick mud with icy cold water about up to the waist.

They struggled out, and Amy, after a change and a hot bath, protested that she never caught cold, and was none the worse. She was urged to postpone her departure again, but Christopher must not be disappointed once more, for she was to join him at Bath next day, since the papers announced the arrival there of some interesting people. But she had a bad shivering fit on the way up to London, and it was evident that she had broken her rule for once and caught a severe chill. She was well enough next day to write an amazing quantity of postcards to her friends, asking them all to come and see her, but not well enough to travel. The day after she was not well enough to do anything at all except to have a high temperature, and all the friends had to be put off.

She grew rapidly worse: pleurisy set in. She became slightly delirious and babbled in a way that puzzled her nurse about garters and strawberry-leaves. It was in vain that she was assured that her garters were quite safe, and when her nurse told her that strawberries were out of season, she said drowsily, 'Yes, but strawberry-leaves aren't.' In the intervals of delirium, though her breathing was difficult, she seemed extraordinarily content and happy.

Then pleuro-pneumonia developed, and Christopher was sent for. It was not a very severe attack, but there were disquieting symptoms.

She made no effort of any kind to fight and resist: she seemed like one who had attained the goal of earthly ambition, and had no desire left for the accomplishment of which she had the will to live.

'I don't like that symptom,' said the doctor to Christopher after one of his visits. 'The state of her lungs is not such as to warrant our taking—well, a serious view of her condition, though of course pneumonia is always anxious work. Her strength is well maintained, her heart action is quite good, but she must somehow be roused. Go in and sit with her, and try to interest her in things which used to interest her. She mustn't talk, but you try subject after subject, and see if you can't get her to rouse herself.'

He shook hands.

'I shall be back about two o'clock,' he said. 'You mustn't be too anxious yet. She has plenty of vitality if we can only get it to work.'

Christopher went to her room. She was lying quite still, her eyes sometimes open, sometimes shut. She knew him, and smiled faintly.

'Now I've come to sit with you a bit,' he said. 'How do you feel, darling? You mustn't talk, you know; better not to talk. I'll do all the talking. Perhaps you'd like me to read to you.'

She seemed drowsy and very apathetic, but her eyes grew a little more alive at this suggestion.

'Yes, read,' she whispered. 'Good Christopher.'

On the table at the foot of her bed were the book by Proust and the new work on Auto-suggestion.

'Ah, I know what you would like,' he said. 'Something out of that book of Proust's which you took to Doncaster with you. Will you give me that book, nurse? Very interesting, I am sure.'

The invalid's face grew fretful.

'No, not that stupid rubbish,' she whispered.

'Well, shall we try that book on Auto-suggestion, dear?' said Christopher. 'You were very much interested in that. You told me you were going to read it in the train.'

Her forehead furrowed itself into unhappy creases.

'Boring nonsense,' she said. 'How stupid you all are.'

Christopher tried the effect of telling her about Bath, but she took not the smallest interest in Bath. He told her how the telephone-bell had been ringing: everyone who knew she was not well had been enquiring after her, and everyone who didn't know had been asking her to dinner. That uncreased her forehead a little, but still she did not seem to care much, and poor Christopher's heart sank. He realized then what a change there was.

He racked his brains for something more. He felt wretchedly helpless, and the waters of Bath had not purged the acidity from him to such an extent that he could think without superstitious forebodings about those picture postcards of Doncaster Castle which for once she had prematurely put into her book. She had said, too, that perhaps there would never be another book—— He felt he had known then that ill-luck would come of her ill-considered act. And yet how diabolical was the Nemesis that had followed it. Just because she had gummed a few picture postcards in. . . . 'I lunched with a Marquis one day,' he said brokenly.

She gave a little sob.

'No, not Marquises,' she said. 'Not Marquises. Garters and strawberry-leaves.'

The nurse had come to the bedside, and was looking anxiously at her. 'That's what she kept saying night and day,' she said. 'I told her that her garters were all right and it wasn't the season for strawberries, and I suppose she got tired of trying to make me understand. And now she's begun again. Whatever can she mean?'

'Garters and strawberry-leaves,' said Amy faintly.

Christopher crushed his temples in his hands. Some remote association, dim as yet, began to form itself in his mind. It was connected somehow with something Amy had written to him in one of those wonderful letters from Doncaster.

'Your garters, darling?' he said.

'No, his,' said Amy.

'She's wandering,' said the nurse, shaking down a clinical thermometer. 'I hope her temperature isn't going up again.'

Suddenly Christopher sprang up.

'No, she's not wandering,' he cried. 'Oh, why did nobody tell me sooner? I know the sort of thing she means, and we'll get at it. She wanted me to read, too——'

He bent over her.

'About the Duke of Whitby, isn't it, dear?' her asked.

A faint flush came on her pallid cheeks.

'Yes, all about him,' she whispered.

Christopher gave a little squeal of triumph, and ran from the room.

He tore downstairs without a thought of his twinging toe, and came rushing up again, three steps at a time, with her copy of the 'Peerage'. He turned rapidly over the leaves with their copious little pencil-marks, until he came to W, and sat down again by her bed, and read.

'Whitby, Duke of. James Francis Adelbert Charlemayne de Vere, K.G., K.C.M.G., K.C.B., O.M., P.C. Born 1882. Educated at Eton and Christ Church College, Oxford. Late Major in 1st Life Guards, Knight of the Order of the Holy Roman Empire, of the Golden Fleece. Married in 1906——' A happy little sigh came from the bed.

'Ah, that's nice,' said Amy, in stronger tones. 'Go on.'

'Married in 1906,' continued Christopher, 'Frances Elizabeth Plantagenetta, second daughter of 5th Duke of Merionethshire, and has issue: John James Plantagenet, Marquis of Pateley, born 1908; Lady Cynthia Elizabeth Plantagenetta, born 1909. Aunts living—— Would you like to hear about the Aunts, darling?'

Amy turned her face towards him.

'Yes, all,' she whispered, 'and the collaterals. And when you've finished them go on to the Merionethshires.'

Christopher read and read and read. There was no end of Whitby collaterals, and the Merionethshires seemed as the sand of the sea for number. But life was coming back to Amy, her breathing grew less distressed, her temperature declined. Half an hour's solid information about these noble lines was poured out in Christopher's sympathetic voice, and she seemed to grow stronger every moment.

At last it was all done.

'I shall get better now,' she said. 'It was just that I wanted, and nobody would understand. Christopher, you've saved me. I feel hungry, too; I should like a little chicken-broth, and then I think I shall have a nap. Tell everyone I am better and shall soon be well. So happy again!'

Dr Elliott came back at about two o'clock, as he had promised. Amy was sunk in a peaceful, restorative sleep and was smiling as she slumbered. A glance at her and a couple of words with the nurse was enough for the professional eye, and he came downstairs again to Christopher rubbing his hands.

'Well, that's all right,' he said. 'You've done the trick, Mr Bondham. A marked change for the better, and I may say she's turned the corner. How did you manage to rouse her to interest in life again?'

'I read to her a little,' said Christopher modestly.

WHEN GREEK
MEETS GREEK

AMY BONDHAM, though far stronger than most horses, was beginning to feel ill with anxiety, for it was now within three days of the date fixed for Mrs Foxinglove's fancy-dress Elizabethan Fair, and still no invitation had arrived for her from the infamous Theodosia. It could not be that a temporary lapse of memory had caused Theodosia to forget her existence, as she had received plenty of reminders. For the last week Amy had been deluging her with hospitalities: she had asked her to lunch, tea, dinner, and supper, all of which Mrs Foxinglove had refused with regrets. Amy had even so far humbled herself as to get a mutual friend to ask Mrs Foxinglove whether she, Amy, was coming to the Elizabethan Fair, and she had replied, firmly and perhaps ironically, that it was no use trying to get hold of Amy, as she was always solidly engaged for weeks ahead. But as the Foxinglove had asked the people she wanted to secure months ahead, this was a very paltry excuse.

To the ordinary mind, such a speech must have seemed final, but then Amy had not an ordinary mind. She meant to go to the Elizabethan Fair, and what to others spelled 'Defeat' spelled to her 'Try again and harder.' In fact, at this melancholy moment she had just tried again. She had written a sweet little note to dearest Theodosia, asking her to come and dine quite quietly with her and her husband on June 18th, which was the night of the Fair, and the answer had come back that dearest Theodosia was out of town that night. Amy read into that mendacious message contempt and an iron determination not to yield. But then Amy was determined too.

For the moment she had winced when these words came pattering into the telephone in the voice of Mrs Foxinglove's butler, and she had an impulse to give up, to leave town for a day or two till the party was over, and perhaps put in the social columns of the leading journals that she had been unavoidably prevented from going to it. That would be quite true, the unavoidable impediment being the absence of an invitation. But her indomitable spirit revolted from the thought of retreat: to quit London would be equivalent to leaving the enemy in possession of the field, and the thought of that steeled her again.

Certainly Theodosia Foxinglove had behaved atrociously, and Amy resolved never to call her Theodosia again. She and Amy were, so to speak, twin dewdrops, for they both devoted their whole pellucid energies to the aspiring art of social climbing; and when, only a year ago, Mrs Foxinglove had left Chicago, where she found it very difficult to rise, and appeared in London, Amy had done a great deal for her, for she had had two years' start, and was chirping away quite high up, while the Foxinglove was still nowhere. She had constantly invited her to her house, she had introduced her to five members of Parliament, two Earls, a prize-fighter, four distinguished literary people, a film-star, a Bolshevist, and a Marquis. There were many others, too, whom she did not trouble to enumerate to herself, but all these she remembered without an effort. At first Mrs Foxinglove had shown no signs of the cloven-hoof; she had indeed behaved very fairly, and several of the brightest butterflies that today refreshed themselves at Amy's hospitable board, had been netted by her at Mrs Foxinglove's house. That was as it should be; that was part of the code of honour that ought to prevail among climbers, and in fact up to the beginning of this season the two had hunted in couples with most gratifying success. But then, it must be supposed, success had gone to Mrs Foxinglove's head; she began kicking down the ladders which had enabled her to attain eminence, and among these ladders (not it is true, a very lofty one, but one that had most emphatically given her a foothold on the lower branches of the great tree of Social Success) was Amy. Indeed, that telephone message she had just received was more than a mere ignoring of her: it was a definite act of hostility and insult. Mrs Foxinglove had 'made a face' at her when she had replied that she would not be in London on the night which everyone knew was the night of the Elizabethan Fair. Of course,

Amy, when she asked her to dinner then, had known also that she could not come, but that was not the same thing as telling a lie. . . .

The telephone that had conveyed this withering message tinkled again, and Amy sprang to it. The Foxinglove might have seen the error of her ways, and even at this eleventh hour have repented, in which case Amy was prepared to call her Theodosia again. But it was only something about grapefruit, and the thought of eating made Amy feel quite unwell. 'She is a snake,' she thought to herself, 'in Grosvenor Square.' Then she pulled herself together, and sat down to concentrate as to how to get to that large reptile-house on the evening of the 18th. She would not give up, she would not retreat into the country, she would not even pretend that she had been asked and had refused. She would go.

At that very moment Theodosia Foxinglove, having emitted that spurt of malice on Amy through her butler, was also concentrating. Though the Elizabethan Fair had been boomed at staggering expense in the social columns of every important journal in London, and a perfect galaxy of distinguished people had promised to adorn it, it still lacked the crowning splendour of being one at which orders would be worn. She longed, as with burning thirst, to curtsy to somebody in her own house, and at present she had not secured anybody to curtsy to. Quite a little curtsy would do to begin with, but she desperately wanted to bring curtsying into her domestic circle. So there she was inspecting the ballroom of her house, which had been transformed into an Elizabethan market-place, with stalls all round it (where her guests could take little trifles, such as gold matchboxes and turquoise brooches, without paying for them) and a dais of seats at one end, violently concentrating as to how to get hold of a Highness. Though all that was otherwise brightest and best in London was coming to the fair, the Foxinglove had still an empty feeling. . . .

Amy dined quietly at home that night alone with her pink, plump devoted Christopher. There was a party or two she could go on to if she wished, but she really did not feel up to it, for though hundreds of friends would be there, the Foxinglove might be among them, and in any case there would be a good deal of talk about the Elizabethan Fair, and that would make her feel faint. So she stopped at home with her Christopher, kind good Christopher. He knew how madly she longed to go to the Fair, for he took the profoundest pride in her social successes, and was much depressed at the way things were going. He had, in fact, anxiously asked her just now if she had 'managed' it.

'I don't know what you mean by "managing" it,' she said. 'I should like to go: I've never denied that. But if you think I would stir a finger to get asked, you are quite wrong.'

'Well, well,' said Christopher soothingly. 'Then that's that. I see. Proper pride: just so. I only meant, darling, that I knew you could get asked in a minute, if you cared to put your wits a-work.'

'How?' asked Amy eagerly, forgetting that she wouldn't stir a finger.

'Oh, somehow or other,' said he. 'Trust you for not being beaten.'

Amy sighed. This was disappointing, for she hoped that Christopher might have an idea. He did sometimes.

'Well, I'm not going to think about it any more,' said she. 'I shall like to have a quiet evening: I'm sure I get one seldom enough. Tell me the news.'

Christopher was skimming the evening paper. There was some melodramatic news about the franc which interested him, but he instantly turned over to the page that would interest Amy.

'Great to-do at the Flower Show today,' he said. 'The whole world seems to have been there. Princess Isabel opened it . . . dukes and duchesses and delphiniums . . . the Prime Minister, Mrs Foxinglove——'

'She would be,' said Amy, suddenly boiling over again. 'I can see her trying to get introduced to the Princess. How people push and shove!'

'Perhaps she knows her already,' said Christopher.

'Not she! She would have put it in the paper that the Princess was among those who had received an invitation to the Fair. Besides, I always distrust that. To receive an invitation means nothing. I might as well give a party and say that Julius Cæsar had received an invitation. All stuff! Also I know that Theodosia, I mean Mrs Foxinglove, would give one, if not both her eyes, to get her.'

'She plays the violin very finely I'm told—the Princess, I mean,' said Christopher, leading the way gently off the agonizing subject.

'Does she?' asked Amy languidly.

There was silence. Amy, a little exhausted by her outburst, sat with half-closed eyes, miserably conscious that in forty-eight hours from now the Fair would be in full swing, and she not there. 'But what does it matter?' she asked herself, and her heart replied to her that it mattered a great deal. Then suddenly she sprang up: a perfectly wonderful idea had come into her head. Whence or how it came she had no notion: she was content to consider it an inspiration.

'I've got a note to write,' she said, and hurried to her table.

The note seemed difficult to compose. Christopher heard the crumple of two or three sheets consigned to the waste-paper basket. But presently it was done.

'Has the last post gone?' she asked. 'Then it must be taken. Ring the bell, dear.'

Christopher looked over her shoulder as she scribbled 'By hand' in the corner, and saw to whom it was addressed.

'Aha! I knew you would manage something,' he said, 'if you cared to give a thought to it.'

Just one half-hour afterwards Mrs Foxinglove arrived at her own door after a little dinner-party she had been giving at the Splendid, at which she had experienced a snub. One of her thirty-three guests happened to be of the household of Princess Isabel, and though the Foxinglove did not know this guest at all, as he had been brought by somebody else, she thought she saw a chance. But it was no good: Princess Isabel, he knew, was dining out on the 18th, and—here he became slightly apologetic—she seldom if ever went to houses she did not know. The baffled Foxinglove therefore came home in a morose mood.

There was her post lying on the hall-table, and on the top a note just left by hand. She recognized the writing, and wondered at the persistence of Some People. Her first impulse was to tear it up without troubling to open it, but it might be amusing to see what fresh assault the impotent Amy proposed to deliver. So she tore it open, and a moment after sat down, with a gasp of astonishment, on a hard hall-chair with a coronet on the back. The note from impotent Amy ran as follows:

YOUR ROYAL HIGHNESS,
This is just to confirm my telephone, and to say how charmed and honoured I shall be to expect you to dinner on Thursday. And what a treat to know that you will bring your violin! Indeed, I will follow

your Royal Highness's wishes and have no party at all. I am not
'going on' anywhere afterwards; it will be so lovely.

Your Royal Highness's
Most obedient and delighted servant,

AMY BONDHAM.

Foxinglove produced a hard short noise in her throat like a death-
rattle, and then began the dreadful business of concentration again.
What had happened was perfectly clear to her lucid mind. Amy had
enclosed the wrong note in the right envelope, and, by inference, she
had sent the other note to a Royal Highness. The inference was quite
wrong, but the upshot was that a Royal Highness who played the
violin, and was thus at once identified, was dining with Amy quietly
on the night when Foxinglove had told Amy she would be out of
town, but was in reality holding the Elizabethan Fair to which she
was determined Amy should not come. But instantly the longing for
a Royal Highness swamped the determination to exclude Amy, and
without pause she seized the telephone, and rang up that obscure
number in South Kensington. She would eat humble pie, she would
drink the water of affliction, but she must be careful not to let slip
that she knew that Princess Isabel was dining with Amy. How people
pushed and shoved! . . .

There was a long pause, and the lady at the Exchange said she would
call 'them' again. At last—

'Is that my Amy?' cooed Foxinglove. 'You dear thing, how are you?
I've just got home, and now I'm lying grovelling——'

'What are you doing?' asked Amy, who had heard perfectly.

'Grovelling,' said Foxinglove. 'Dust and ashes, can you hear me?
I've made two quite awful mistakes, and I can't think what you'll say
to me. First, my stupid butler told you I was out of town when you
so kindly asked me to dine on the eighteenth, and I thought it was
the nineteenth you said. I'm here all right on the eighteenth, but,
alas! I can't dine as I've got a little party that night.'

'Ah, yes,' said Amy. 'Of course, I quite understand. Such a natural
mistake.'

'And my second mistake is even worse,' said this remarkable liar,
'for I find that my stupid secretary hasn't asked you to it. I can't
make out how it happened. Now do forgive me and come.'

It was Amy's turn to say 'Alas!'

'Alas!' she said, 'there's a friend dining with me that night. Just proposed herself. And we shall be having a little music, as she plays the violin, and I don't really know when——'

'But you won't go on making music till four in the morning,' interrupted Foxinglove. 'Come in after your old tunes. And bring your friend. Always delighted to see any friend of my Amy's. Who is your friend?'

Amy's irritating laughter tickled Foxinglove's ear.

'Oh, just a friend,' she said. 'I'm sure you don't know her.'

That was a nasty one, and Foxinglove winced.

'Well, bring her along for an hour,' she repeated. 'But anyhow, come yourself. Promise! And your friend, too: ever so welcome. Ring her up tomorrow, whoever she is, and say how pleased I shall be: Elizabethan Fair: fancy-dress.'

'Sweet of you,' said Amy.

'And forgive my secretary's mistake, dear,' said Foxinglove.

'Why, of course,' said Amy genially. 'But you'll forgive me, won't you, if I can't manage to look in? If it's fancy dress, I'm afraid——'

'Fancy-dress optional,' said Foxinglove. 'And indeed I shan't forgive you and your friend if you don't come. I shall feel real bad about it.'

Amy managed to go to the Elizabethan Fair, for she had her fancy-dress all ready in case. She arrived quite alone, rather late. The galaxy of fashion which had assembled for the Fair was sitting watching a Morris dance of highly decorative Elizabethan yokels. The Foxinglove, whom nobody could mistake for anyone but Queen Elizabeth, hurried towards the door with all her pearls a-jingle as her name was bawled out, and they kissed affectionately.

'And your friend?' asked Foxinglove.

'Couldn't persuade her,' said Amy.

DOGGIES

IT had really been difficult to know what to do with this sumptuous access of fortune, and Amy Bondham, whose very eccentric uncle had left her a couple of hundred thousand pounds, almost wished that it had been less. She and her Christopher were childless, there was nobody whom either of them wished to make rich, and her imagination, usually so vivid, could not figure how to spend an additional eight thousand a year.

She and Christopher were dining alone on the night after the eccentric uncle's cremation. It had been an agitating day, for the deceased had left a very odd will which included the strange provision that on pain of forfeit of this substantial fortune, Amy, dressed in white, should directly after the cremation scatter his ashes in Piccadilly Circus with ceremonious gestures, and she had found it very trying. Christopher, however, had suggested that nothing in the will prevented her making her ceremonious gestures from the seclusion of a taxi, and then throwing all that was incombustible of her uncle out of the window, and she had done this in the presence of her uncle's solicitor, who sat beside her. The ashes had blown in the face of the policeman at the crossing, and set him sneezing: it had been a nerve-racking performance. . . . But it was over now, and this evening she and her husband were wondering, with anxious faces, not how to make both ends meet, but rather how to make them not meet, and thus employ in a self-respecting manner this large sum of money. They had already all that they could conceivably need to oil the wheels of their passage through this vale of woe, and a means of rational expenditure yielding a high dividend of enjoyment and advantage was what they sought for. Amy's sole objective was social success, and Christopher racked his brains for suitable suggestions.

'You might like a bigger house, dear,' he said. 'There's a charming house in Berkeley Square of which the lease is for sale. I had a look at it this afternoon. Fine, big reception-rooms, a dining-room where you could seat fifty——'

Amy shook her head.

'No. I don't want that sort of house,' she said. 'Those great entertainments only help people who want to get on in a vulgar, blatant manner. Far more chic to have our little house here and not collect mobs.'

She sighed.

'I used to think how wonderful it would be to have huge dinner parties and great balls,' she said, 'but really that is out of date. A dozen people to dinner, as long as they are the right ones, is far more telling. The other's not my line. . . . Then I had thought of endowing some charitable institution, a hospital or something, but what should we *get* by doing that? You could be made a peer or at least a baronet, because whatever they may say, honours are purchasable, but that wouldn't make me a penny happier. It's far more distinguished to be Mrs Bondham than to be Lady Bondham, especially if everybody knows why.'

Christopher ruminated over this. In his secret heart he would have liked to be Sir Christopher (at least), but if it wouldn't give Amy any pleasure there was the end of it.

An idea struck him: if Amy didn't want a bigger house in London, to which her amazing and hospitable activities had hitherto been completely confined, there were places outside London which gave scope for social expansion. August and September, which were slack times in town, were already adequately filled, for there was a month at Aix-les-Bains, and visits to country houses. But she had always found January and February rather unoccupied, and though she professed to be delighted to have a few quiet weeks to herself, he was sure she would sooner share them with other people.

'A yacht,' he suggested brilliantly. 'Cruising in the Mediterranean: Mrs Bondham with a few friends on board.'

Amy looked at him in amazement.

'Darling, how can you be so silly?' she said. 'As if you don't know that I'm the worst sailor in the world.'

'But I'm not so silly,' said he. 'I don't suggest you should cruise except where it's quite flat. There are harbours in the Mediterranean, I am told. You could go overland and join your yacht at Monte Carlo, and have no end of friends on board. Then you could take the train and join it again at Naples or Taormina.'

She turned her head to him.

'You're not silly: I withdraw that,' she said. 'Go on: let me hear more.'

Christopher gave her a very bright little sketch. The idea seemed most attractive to him personally, because he adored the sea, and hated the climate of London in winter: two or three months in the Mediterranean were much more enticing than two or three bronchial colds in Mayfair, and he figured himself in a peaked cap, a black coat with brass buttons, and lawn-tennis shoes. Naturally he did not lay stress on that, but indicated how Amy would come out overland to Monte Carlo, and there find her beautiful yacht lying in the harbour, how she would collect friends who would dine on board and dance on deck, and then rejoin it again, unless the sea was perfectly flat, at Naples. The illustrated papers would be full of her yacht.

His little picture kindled her imagination, but not, unfortunately, on his lines. Certainly he was right about getting out of London during those months, and there was something pleasantly sumptuous in his scheme. But there were objections, for supposing the Mediterranean should choose not to be flat, she would spend her winter in making land journeys up and down the coast in order to dine occasionally in a harbour. It hardly seemed worth while. And in the intervals was Christopher to be careering about at sea with entrancing guests on board? She quite trusted Christopher of course, but still——

'Well, that is an idea,' she said, 'and it's clever of you to think of it. It certainly has its points, oh, certainly, and we must bear it in mind. What we really want is to see delightful people and do delightful things instead of living under a blanket of fog in an empty town. But the Monte Carlo crowd . . . you know they are not really our sort. Nothing but gambling all night and dressing-up all day. I think you and I would feel starved there.'

'But all the world goes there in the winter,' said Christopher, who yearned after his yachting-cap.

'Indeed it does not, for the majority of my greatest friends, the

ones I really value, go down to their places in the country, and shoot
or hunt. The Harrogates, the Middlesexes, the Pateleys, the
Bidefords, they all spend those months in Leicestershire, and we see
nothing of them all the winter.'

She jumped up.

'Oh, Christopher, I think that is what we must do,' she said. 'A
box—I notice that in Leicestershire they call houses boxes—I should
really like to take a nice big box, somewhere down there this winter,
and see more of all those dears. It would be a new side of life: I
often feel when I'm with hunting-people that there's a big piece of
them of which I know nothing.'

'But I should be expected to hunt,' said Christopher in some
dismay. 'So would you.'

'Well, I don't see why you shouldn't hunt,' said she. 'Anyhow, you
can see what it's like first, and then if you feel you can't manage it,
you can easily get out of it. If there's a meeting of the hunt—don't
they call it a meeting?—anywhere close, you can have a cold or be
obliged to go up to London. As for me, I shall certainly say that I
don't hunt, but why should that cut me off from all my friends?
Where's *Country Life*? I've often noticed that there are hunting-
boxes advertised in it. We must look, and try to find a box
somewhere near the Bidefords. We shouldn't want any shooting of
course, for I know you would hate to be obliged to shoot, and I
rather think that where there is hunting there isn't shooting. And
there are Hunt-balls: I believe they have the greatest fun in
Leicestershire. We will go into it all.'

Luck always attended Amy, or perhaps it would be truer to say
that her determination, when once she had set her mind on
anything, had something of the compelling power of faith about it,
and made to happen that which she wanted to happen. They set off
to Aix a day or two afterwards, and the very first person Amy ran up
against was Lady Bideford, who told her of precisely the house that
seemed likely to suit her. It was in the middle of the Paston country
(of which famous pack her husband was Master), and he would be
delighted to know that somebody friendly to hunting would be
occupying it, since the owner, who was going abroad for the winter,
was a wretched curmudgeon without a spark of sportsmanship in
him, and would never allow the covers in his park to be drawn.

'It's the one black spot in the county,' she said. 'The horrible man refuses to let us hunt there, and I believe the place is crawling with foxes. You'll be a public benefactor, and we'll put up your statue, and have a meet there as soon as ever you get in.'

Amy noticed the word 'meet', which was evidently the correct version of 'meeting', and got the address of the house-agent.

'I do hope I shall be able to get it,' she said. 'I shan't hunt myself, but it will be so delicious to be among friends when London is empty in those dull months. Christopher is looking forward to it so. And I'm so fond of dogs.'

Lady Bideford naturally supposed that Amy had suddenly changed the subject when she said she was fond of dogs, for it never occurred to her that she could be alluding to hounds as dogs. They weren't dogs: they were hounds.

Amy's application to the house-agent was successful, and though the rent asked for Cold Bovington House was rather high, as being in the centre of the best hunting-country, it did not make a very large hole in the newly bequeathed fortune. The house was big, but pleasant and cosy, and though the country round, of clayey soil, was extremely dreary, consisting, as it did, of flat grass fields intersected by ditches and thick bare hedges, the Harrogates and the Middlesexes and the Pateleys and the Bidefords were all within very easy distance, and Amy was soon asking them all to lunch to meet each other in her urban style. But this was rather a disappointment; she had supposed that people hunted in the afternoon just as they played golf and lawn-tennis in the afternoon, but it appeared that they hunted in the morning, and continued doing so as long as it was light, and so were practically never back to lunch even if it rained, for hunting went on in the rain exactly as if it was fine. She gathered, however, that if it froze hard, and there was bone in the ground (whatever that might mean), hunting stopped. So she hoped it would soon freeze.

Christopher had subscribed liberally to the Paston, and had received a card which gave him a list of its fixtures. This was rather embarrassing, for though the ostensible reason for the Bondhams having taken Cold Bovington was that he should get some hunting, one glance at a run which swept across those fields with their high hedges and deep ditches on the morning after they arrived was conclusive as far as his hunting went. Amy quite concurred.

'My dear, it looks terribly dangerous,' she said. 'Oh, there's a horse fallen down and the man's come off. How awful! I wonder if he's hurt. You must promise me not to attempt it: I had no idea it was like that.'

Christopher hastened to relieve her mind, but when the list of fixtures arrived, it was clear that a tangled web must be woven without delay. The Paston met with terrible frequency within easy range of Cold Bovington, and on the approach of these horrid mornings, Christopher had to arrange to be summoned up to London on business of high importance.

Amy did her part, when she motored off to see the meeting (which she almost always remembered now to call 'meet') and explained to the Duchess of Harrogate and Ladies Middlesex, Pateley, and Bideford that her husband (so disappointed) had been obliged to go up to London. That was easy for Amy and productive of pleasant conversation, but meantime poor Christopher was spending four hours and more in the train, lunching at his club, passing the idle afternoon in naps or cinemas, and arriving back at Cold Bovington in the evening, bilious with his sedentary day. To be sure it was better than hunting, but that was all that could be said for it.

Sunday alone was a perfectly safe day for him at Cold Bovington because there was no hunting then, and Amy had an opportunity of getting the Harrogates and the Middlesexes and the Pateleys and the Bidefords over to lunch. Christopher could then lament those annoying summonses to London which had prevented him taking part in the wonderful runs they had been having.

On the second Sunday of their tenancy, there was a dinner to which they were bidden at Lady Bideford's. Christopher had been delighted to observe the very marked fall in temperature which had occurred during the afternoon: the wind had shifted to the north, and he had read with strong satisfaction in the Sunday paper that there had been a heavy fall of snow in Derbyshire, and the general weather outlook was unpromising. Snow, in moderate quantities, so Amy told him, interrupted hunting as effectually as frost, and there really seemed a good hope that he would not be called up to London next day. But if there was any question about it, he would certainly have to go, for the meet next morning was at his own house. At the end of dinner Lord Bideford moved up next him with port and cigarettes, and expected great things.

'We've been stopping the earths,' so Christopher understood him to say, 'in the big cover that comes close up to your house, and we shall find there for a dead certainty. These last three years we've never been allowed to draw your covers, and we all much appreciate the fact that we've got a good sportsman there now. We'll have a rare day tomorrow if there's no snow. You'll be out with us, of course. You've not had a day with us yet, have you?'

Christopher fortified himself with a glass of port and the recollection of the very unfavourable weather forecast.

'No, I've had rotten bad luck,' he said. 'Every day that you've met since I've been down here, I've had to go to town on some annoying business. But I hope I shan't have any summons tomorrow, and get a day's hunting instead. No sport like it, is there? What time will you meet tomorrow?'

'Ten o'clock,' said Lord Bideford. 'I should like to have made it nine, but there are a lot of lazy folks. We'll meet right at your front door, and draw the big cover alongside straight away.'

Christopher knew the morning trains by now: he would have to catch that melancholy 8.15 a.m., unless kindly snow or frost came to the rescue.

'Splendid!' he said, hearing the faint unmistakable patter of snow at the window. 'I trust the weather will hold up. But it's bitterly cold: looks like frost or a big snowfall.'

It was still cold when, after a rather expensive rubber or two at bridge, he and Amy went home. The sky unfortunately was clear now and the prospect of an effective fall of snow was remote. But, on the other hand, a clear sky might mean frost before morning, which, if reasonably severe, would do just as well, for the ground was already lightly frozen, and an adequate supply of 'bone' might develop during the night, thus allowing him to have a quiet day in the country. He went to bed in hopeful mood, but woke in the small hours, to hear, to his dismay, the fatal sound of rain on his windows: you could never count on this wretched climate. Morning broke dark and windy and warm, and it was necessary to catch the 8.15. As his train crawled London-wards, the rain ceased and the sun came out, and he imagined to himself how delicious it must be in the Mediterranean, where no foxes had holes, and no packs of dogs would be hounding him from the deck of a commodious yacht.

By half-past nine the animated scene began at Cold Bovington. Lord Bideford arrived early in a scarlet coat which Amy now knew to be of the shade called pink, and presently the meet assembled. Amy kept popping in and out of the house, enjoying herself immensely, for Harrogates and Middlesexes and Pateleys with wives and sons and daughters and horses and motor-cars looked in, and she found everybody so friendly and full of cordiality at this admission to covers so long shut to them, that she felt that she was indeed the provider of this rich entertainment. And would she not think about buying Cold Bovington House? The old unsportsmanlike owner, Lady Middlesex believed, was quite ready to sell it, and Lady Pateley had heard that he was never likely to live in England again: lungs, my dear. Then the Duchess of Harrogate, who had not had any breakfast, came in and had a slice of cold ham and a cup of tea; in fact, there could not have been a more wonderful morning picnic of the best and brightest. The hounds were gathered outside, and the whips, whom Amy had learned not to call ostlers, were preventing them from invading flower-beds, and there they were in a compact, good-tempered mass, wagging their tails (Amy had not learned 'sterns' yet), and nothing could have been more English and sporting and aristocratic. Then Lady Pateley hoped she would dine with them tomorrow night (and of course Christopher), to meet a Royal Highness who was staying with them for two days' hunting, and finally Lord Bideford said that he had given everybody ten minutes' law, and so he wasn't going to give them any more.

He went out to mount, and Amy and the picnic followed. Outside there was a perfect mob of horses and grooms and motor-cars and all those nice dogs whom the whips knew individually by name. In five minutes they would all have moved off, when, at the very last moment of her triumph, Amy, suddenly remembering how much she liked dogs, had the most ill-inspired and incredible idea. And yet it seemed such a suitable dog-loving thing to do. . . . She ran back into the dining-room and, rummaging in a cupboard of the sideboard, found a tin of small sweet biscuits. With this in her hand she tripped out again.

'Oh, do wait one minute, Lord Bideford,' she cried, 'and let me give your dear doggies a biscuit each. What pets!'

For one moment it was as if the whole of the Paston hunt was turned to stone. The next it burst out into shrieks of inextinguishable laughter: whips and ostlers and riders all yelled. Twice Lord Bideford failed to command his voice before he could utter a word.

'So kind of you, Mrs Bondham,' he said, 'but I am afraid the doggies are in training. We're very strict with them.'

The hunt moved off: every now and then a peal of laughter broke out again, and presently Amy was left alone at the front-door with her unopened biscuit-tin. Her kindly thought for the doggies, she perceived, must have a comic side to it, and she had no idea what it was.

The story ran through the county like fire through dry stubble, leaving in its track not desolation but hilarious mirth. In a couple of days everybody knew it and it must have reached the ears of Lord Bideford's distinguished guest before the dinner, for when Amy was presented to him and told him how the Paston had met at Cold Bovington House, he suddenly burst into giggles. Somehow she had become a comic: people didn't want to laugh, but they couldn't help it, and she could not but perceive that nobody mentioned hunting to her any more, while if she initiated the topic which so largely interested them, they gave faint grins and firmly changed the subject. She began to perceive that she was getting no nearer the real life of hunting-people: she felt she was branded, like some kindly innocent Cain. The place suited Christopher no better, for the weather remained deplorably mild, and he had to go up to London about four times a week.

He had come back one evening after one of these objectless excursions, tired and dejected, for it really was a doggy's life. He had brought the evening paper down with him, and Amy, idly turning over the leaves, read a paragraph or two about the sparkling sunshine on the Riviera, and the enormous fun that no end of smart people were having there. She glanced through that with a certain interest, for she knew a great many of those distinguished triflers, and then her eye fell on an advertisement beginning, 'SEA-SICKNESS A THING OF THE PAST. . . . This perfectly harmless and infallible remedy is vouched for by . . .'

She read it through, and laying the paper down, glanced at Christopher. He was sitting close by the fire, for he had a bad cold, and was stertorously dozing. In a moment her mind was made up.

'Wake up, dear,' she said. 'We must have a talk. How depressing this place is: why did we ever come here? I want the sun, and so do you. And this advertisement: look at all the testimonials! A bishop, a prize-fighter, an Earl. Let us think about yachts.'

CRANK
STORIES

THE CASE OF
BERTRAM PORTER

BERTRAM PORTER was barely thirty when he died in the meanly tragic manner which will be here narrated, but for years before that he had been a very great nuisance to all those who had the misfortune of being acquainted with him and who had not dropped him. In all respects but one he was a delightful fellow, full of kindliness and humour, cheerful, unselfish, indulgent to the weakness of others, and elastic under the blows and buffets which this remarkable state of affairs called human life rains down upon the most fortunate of our race. But this one defect of his was serious. It was as if a homicidal maniac was recommended to our good graces, and we were told that in all respects but one he was a man eminently livable with. Also, in the sad case of Bertram Porter, his one defect gradually invaded and blotted out his other merits, even as the cloud, no bigger than a man's hand, effaces the blue sky of a benignant summer day. His defect—to put the reader out of suspense, and to assure him that no attempt is being made to interest him in the career of some moral delinquent like a cannibal or a second-rate pianist—was that he thought about his own health.

When first I knew him, he was a medical student at one of the great London hospitals, and he appeared to be a person who might quite easily become of some use to the world and a pleasure to his friends. But even then the little cloud was already risen on the horizon, and I became aware of it when he dropped into my rooms late one night to smoke a pipe and drink a glass of whisky and water.

'You haven't got such a thing as a piece of sticking-plaster?' he asked as he entered.

I had not, and told him so.

'I scratched my hand somewhere getting down from the 'bus,' he said. 'No, it is not exactly bleeding, but the outer part of the skin is rubbed. Ah, well, it doesn't matter! But have you any sort of disinfectant I can dab on it?'

I was not the fortunate possessor of any disinfectant, but suggested that, if he were anxious, the fire was burning well, and a red-hot poker could easily furnish him with a satisfactory cautery. He laughed at this, lit his pipe, and helped himself liberally to whisky and cold water. It seems to me now quite extraordinary that it is only eight years ago that I saw him freely partaking of these two violent poisons—alcohol and nicotine. Then he settled himself in his chair, and began talking in that eager manner that was so characteristic of him.

'Yes, all you fellows who know nothing about your wonderful bodies may laugh at me,' he said, 'but the more I see of illness in the hospital, the more I learn how preventable it all is. There is not a single disease that we could not prevent if only we all knew just the elements of the laws of health, and took the simplest precautionary methods.'

'Typhoid and cholera, and that sort of thing?' I asked.

'Oh, not only those,' he said. 'Cholera is stamped out in England, typhoid soon will be, diphtheria is no longer to be feared now we have the antitoxin treatment. There are the great foreign enemies, so to speak, that attack us, just as this scratch on my hand is for the moment a place where a foreign enemy, a staphylococcus or streptococcus, might enter.'

'Don't know them,' said I.

'Lucky fellow! If ever you had had a boil or a carbuncle, you would have made acquaintance with a staphylococcus; if ever you had had an abscess, you would have harboured a streptococcus.'

'I did last year. I suppose the dentist took the streptococcus away. That will do, I suppose?'

'Yes, and he took the tooth with it. You ought to have been in such a condition that your phagocytes ate up the streptococcus, instead of letting it breed and multiply.'

I cannot now remember all he said that night, for he talked late, and indulged more than once in the two virulent poisons of which I have spoken—alcohol and nicotine—and his discourse was somehow strangely fascinating. He told me about the beneficent armies of phagocytes that swarm in our blood, ready to pounce on and devour all malignant intruders, about the means whereby they can be encouraged to eat the various microbes which from time to time find lodgement within us. He spoke, too, not only of the external foes that assail us, but of the auto-intoxicants that are strangely bred within us, foes of our own household, so to speak—the self-generated poisons that produce gout and rheumatism and arterio-sclerosis, diabetes, possibly cancer, and all the diseases which age and kill people when the first vigour of their youth is past. All these, he said—far more insidious foes than the hordes of typhoid- and cholera-producing germs—could certainly be met, and, what was more to the point, be guarded against, so that a man might and should attain his century of years and more in serene and splendid health. As I have said, it was all intensely interesting, and yet at the end, when he rose to go, and lit yet another pipe to comfort him on his way home—for all vehicular traffic had long ago ceased— I felt somehow distrustful of it all.

'Well, it's all like a charming fairy tale,' I said, 'and I shall try to think of it as a fairy tale, if I think of it all. But probably I shall do my best to forget it.'

'But why?' he asked.

'Because if I thought of it as real, which I am sure that it is, I should devote all my time to thwarting the enemies and breeding the friends. I dare say it's all right for you, since it is your profession, but it would never do for a man who was employed in other ways. I shall leave my—my opsonic equation to solve itself. Anyhow, I am glad to see you are not consistent yourself. You have told me that nicotine is a cardiac irritant, and thus disturbs the circulation, but if that isn't the fifth pipe you have smoked tonight, it is the sixth.'

He looked at it regretfully—it was burning well—and then knocked it out on my hearth. 'You're quite right,' he said. 'I smoke too much. Also, nicotine is an irritant of the mucous membrane, and certainly makes it more liable to harbour germs.'

'But it is a disinfectant, too,' I said.

He laughed.

'By Jove, I wish you had reminded me of that first!' he said. 'Well, good night!'

I always look back on that evening as the first on which I became aware that for poor Bertram Porter a cloud, a mental storm which should overlay the sky, was within view. For a while it remained on the horizon, for he was busy with his work, and devoted his energies in the main to making other people well, instead of keeping himself so. He did not, at any rate, at once give up the irritant poisons of daily life, but, on the other hand, the fact that he was liable to go into houses where there was neither sticking-plaster nor disinfectants made him careful to carry these about with him. It was not long after that we spent a few days in the country together, and, walking through a wood, I carelessly let a branch of bramble fly back against him, so that his hand was punctured by six or seven infinitesimal wounds. He was not in the least annoyed with me, but he stopped at once and took from his pocket a little phial of disinfectant, and dabbed a drop on each of the punctures. Then, with his deft doctor's fingers, he cut seven minute squares of black plaster and covered the pricks with them. This was done almost automatically, for he continued all the time to talk about the atrocious golf he had played that morning. There was no gainsaying the atrocity of it, but I would cheerfully have assented that it was the most remarkable example of professional skill in an amateur that I had ever seen if I had known what lesson he was going to read from it.

'Of course, all games requiring delicacy of touch and accurate timing,' he said, 'are not what they call "mere games" at all. They demand the most accurate adjustment between various nerves and muscles. The eye has to notice the exact spot a ball occupies, or, in the case of a moving ball or at lawn tennis, will occupy, a fraction of a second after it is seen, and the muscles of the arms have to act in unconscious obedience to the eye. The motor and sensory nerves have to synchronize, if you see what I mean.'

He paused a moment and adjusted the last piece of plaster.

'There are various causes that may give rise to an imperfect synchronization,' he said, 'so that one times the ball wrong, as I did once or twice this morning. It may arise from fatigue of the nerves, or fatigue of the muscles, or, again, an act of indigestion, so small that you are not conscious of it, may put the mechanism out of perfect order, or some inequality of circulation.'

'It is mostly due to taking your eye off the ball,' I said.

'Certainly. But why do you take your eye off the ball? Because the muscles that govern the movement of the eye are acting imperfectly, and because the muscles of the arms are slower to act than the eye knows, so that you have not actually struck the ball when your eye tells you that the stroke ought to have occurred.'

It was clear that even the innocent game of golf was being made grist for the mill of health, and I tried to rescue it.

'I take my eye off it,' I said, 'merely because I am not attending, but thinking about something else. There is nothing so fatal as to try to reduce games to microbes and nerves. You shall eat at lunch today all that you know is most digestible, you shall rest afterwards, if you like, you shall convince yourself that your heart is acting normally, and yet your game will probably be below contempt. And the reason will be that you can't play golf. There is no reason why you should. I can't play it, either, but I am quite well.'

He sighed.

'Still, there must be a reason,' he said.

I have detailed this trivial conversation at length because it marked a stage in the development of poor Porter's disease, which here seemed to have invaded the realm of recreation, and to be no longer confined to his work. And very shortly after, the determining cause of his ruin came in the apparently cheerful fact that an aged and ailing aunt of his died, leaving him a fortune of a hundred thousand pounds. He instantly retired from his profession and devoted himself to what he called 'original work'. Money had given him leisure, which to some natures is the greatest blessing, but to most the greatest curse. To him it proved the most pungent of curses, for it made it possible for him to study the laws of health and life, not for the sake of alleviating disease and pain in others, but by concentrating his whole mind on himself.

He began by leaving London and settling in an odious villa outside Brighton, in order to enjoy a greater purity of air. This he learned about by careful analysis, and found that a sample of the Brighton atmosphere, even in the centre of the town, contained twenty per cent fewer microbes than a corresponding specimen taken from Berkeley Square. This 'ampler ether' he specially inhaled in bouts of deep breathing, and I have often come upon him, in a corner of the garden, with an expression of one about to burst, or else snorting like a steam-engine. He gave up alcohol altogether, and took to drinking sterilized milk with his meals. The sterilized milk led towards obesity, so he made a practice of running aimlessly about for an hour and a half before breakfast in order to counteract this. The running about led to exhaustion in the middle of the running, so he instituted a period of rest from eleven to twelve in a recumbent position, and a small refection of minced beef in a brown bread sandwich. Also, he cut off the use of tobacco, which led to a certain instability of the nerves, or so he supposed. But that was easily remedied by occasional doses of bromide of potassium. That, it must be understood, was only a temporary measure. He soon got not to miss his tobacco, and therefore abandoned the bromide. But I think he rather missed the bromide.

Simultaneously the question of diet occupied him. It was rather an elaborate affair to cut off all alcohol, substitute milk, and then be obliged to run about in order to avoid the consequences of milk, so he took to some temperance and fizzy drink, which made him bilious, and corrected the biliousness by some sort of pill—I think podophyllin. Flesh foods of all kinds were, after a year or two, thrown overboard, and he lived mainly on brown bread, cheese, fruit, and nuts, which, after a prolonged trial, he settled were the best sort of food. Very soon he took to consulting a little table of food values, and, whether hungry or not, ate the same quantity at every meal. The quantity was weighed on a neat pair of scales which he put beside him on the festive board.

All this and much more of an intricate and complex regime was but the preliminary step of putting himself in as good a condition of health as possible in order to enable him to do his work in the most efficient manner, and settle the question of microbes. Anything that could possibly excite him he also kept far from him, since excitement led to irregular action of the heart and sleeplessness at night. All games, therefore, including bridge, were barred, since one night he lay awake thinking of what would have happened if he had finessed his queen, and he would not even play golf. I had the pleasure of pointing out to him that, since now he was living a life of absolute salubrity, he would always hit the ball exactly in the manner he meant, and would go round any links in sixty or so, but he thought best not to make the experiment. Incidentally, also, he fell in love with a most delightful girl, but, as will be easily conjectured, he found her most disturbing to the nerves and circulation, and chased her from his emotions as if she had been a streptococcus. Bromide had to be called in again for a time, but he was soon himself again, and was now ready to begin work.

About two years later his house was struck by lightning, an invasion which he had not contemplated. This necessitated the erection of an elaborate system of lightning conductors, and a new dome-like roof to his house. The lightning conductors, four in number, soared high as masts separate from the house, so that the most powerful of flashes could not but be attracted to some one of these sublime spikes, and would thus be promptly earthed. Meanwhile, his apparatuses for the detection and destruction of microbes continued to arrive in large packing-cases, and the house may be said to have become a microscope. By degrees, as his work progressed, and he began to learn more of the insidious ways of microbes, it also became more and more uninhabitable, and himself more and more impossible. I was living near Brighton at the time, and there saw a good deal of him, and was constantly shocked to observe how ill he looked. His organs, he assured me, were all perfectly sound. He also assured me that he felt in supreme health, and we must suppose that the incessant worry about the possible invasion of his dwindling frame by some chance bacillus so preyed on his nerves as to produce the appearance of a man far gone in consumption. He grew thinner and thinner, but was delighted with his emaciation, since all adipose tissue, he informed me, was perfectly useless, except for purposes of warmth, and was merely a sort of cucumber frame or hothouse for the foe. So he obtained the necessary warmth by putting on extra thick sterilized underclothing, and heating the house till the head of the ordinary individual swam with dizziness. Then, gradually he became convinced that even the air of Brighton was not good enough for the lungs of the person who wished to be really healthy, and all windows were hermetically sealed, and the house supplied with oxygen by a chemical process. In every room was a pipe-fed apparatus which panted oxygen into the enfeebled air, and an oxygen gauge, so that, when the air was adequately vital, you could temporarily turn off the supply. If you forgot to do this, a sort of heady drunkenness, a fictitious exhilaration, seized the occupant; and more than once he and I, engaged on some discussion, have had to stagger from the house and lie down on the lawn outside, running the risk of harbouring Brighton microbes. Exercise in the open air was similarly abandoned, and our poor friend, whom we can no longer consider sane, used to get into a sort of clothes-basket, which, when he turned on an electric current, performed a kind of St Vitus's dance, which gave those who sat in it all the benefits of horse exercise with no risk except that of being tipped on to the floor, where you fell on a well-stuffed and sterilized mattress. The bed linen was made of some vegetable product, and smelled faintly of carbolic, and in the morning, after a bath of sterilized water warmed to blood heat, Bertram Porter sprayed himself with a weak solution of Salol.

It was something, however, to learn from him that all these precautionary measures were temporary also, like the bromide. His plan was to render himself immune to the attacks of all hostile microbes, while diet and a copious consumption of the famous Bulgarian bacillus would guard him against the probability of auto-intoxication. When completely immune, his intention was to offer himself as a subject to the doctors, who regarded him as a crank, and let them inject typhoid and diphtheria and anything else they pleased. Naturally—this was the golden dream—all mankind, following his example, would, like him, stand beyond fear of every sort of disease, and live—he could not say how long they would live—but certainly a man of a hundred would still be in the vigour of his youth. But, to attain this end, it was necessary for him first of all to demonstrate this impregnability of his own person against all malignant microbes—in other words, he had to give his phagocytes omnivorous appetites, so that they would instantly and greedily devour anything hostile submitted to their notice. Till then it was but the part of prudence to erect those temporary barricades in the shape of sterilized water and antiseptic sheets against their chance onslaughts. These were but the scaffold. When the impregnable house was finished, down the scaffold would come, and Bertram Porter would bathe in diphtheria and drink typhoid neat, and brush his hair with a comb impregnated with scarlet fever.

Till then there were no carpets in the house, for they harboured dust, and it seemed to me, sitting in a stiff and shiny chair, that the very sunbeams lacked the dancing motes that so gaily play there in a less barren air. That morning, I remember, Bertram Porter had his head completely shaved, because hair was a harbourer of microbes, and gave away a Persian kitten because it would go out of doors and no doubt pick up all sorts of infectious things. He had tried brushing it with carbolic powder, but there was no certainty of destroying all hostile life; also, the kitten showed a desire to lick the obnoxious stuff off its soft blue coat, and incurred a risk of carbolic poisoning. He sat opposite to me at a bare deal table, with his eye glued to his microscope, his ears stuffed with antiseptic wool, and his scalp gleaming, weird and white, in the moteless sunshine. He had told me that there were seventy-nine microbes which were hostile to existence, and when once a man was immune to them, there was really no reason why he should not consider himself immortal as long as he kept away from motor-cars.

'And do you propose to inoculate yourself with all those?' I said. 'Is life really worth living if you are to have every disease under the sun, though only in modified form?'

He did not answer for a moment. Then he withdrew his eye from his microscope with a sigh.

'That was most interesting,' he said. 'A very weak solution of chlorine destroyed the common bacillus of typhoid in nine seconds. Inoculate myself with seventy-nine diseases, did you say? Certainly not. Some fifteen will be all that are necessary—the more virulent sorts. I went through the typhoid inoculation a few weeks ago. Dear me, I really wish you would let me do the same for you! It is just the injection of these few dead bacilli, and I stake my professional reputation that you will have no more than a day or two of malaise. Think of the risk you daily run of typhoid!'

I had to shout at him—he was plugged up with antiseptic wool— also shouting was suitable for emotional reasons.

'But that is just what I won't do!' I yelled. 'I will not think of the risks I run. I would sooner die tomorrow than live a hundred years on your terms!'

He looked pained, and unplugged one ear.

'But you don't understand,' he said. 'By the time I am thirty-five I shall be immune from all known diseases, and have at least seventy years of secure vigour before me.'

'Do you call yourself vigorous now?' I asked.

'I am in perfect health,' he said, 'and have been so ever since I began to exercise ordinary care in my mode of living. You, on the other hand, have had influenza twice in the last four years. Dear me, it is nearly lunch-time! I will have five minutes in the exerciser first.'

'I would sooner be a Christian Scientist, my dear Porter, than I would be you,' I said. 'I may have had influenza, but I have had some fun.'

By virtue of my blindness and unreasonableness, I was allowed proper food and drink at lunch, while he with meticulous care weighed out his nuts and beans. Just now he ate a good deal of fresh fruit, for the sake of some salt contained in it, and today, though the month was still early April, a small basket of cherries appeared at the table. I ate two or three, but they were perfectly tasteless, and Bertram Porter finished them, washing each in a glass of water and drying it afterwards.

'There, again, is one of the most elementary precautions,' he said, 'which I am sorry to see you neglect. The damp skins of fresh fruit are a regular grazing ground for microbes. These cherries must have come from the South, and who can tell through what tainted and infected air they may have passed? I saw, for instance, that there was cholera at Marseilles. It is highly probable that our cherries came from the Riviera and passed through the town.'

'And you haven't inoculated yourself for cholera yet?' I asked. 'How careless!'

'No, I am not sure that I should, even if cholera came to England. The experiments at present made have been far from reassuring. The bacillus is peculiarly virulent, and certainly at present I should not think of being inoculated. Won't you have any more cherries?'

He went on eating them, and, as he ate, launched out into one of his extraordinary disquisitions about the tumult and battle that goes on hourly within us. Tragic as I found this practice, his theories were always fascinating, and he became positively epic over these microscopic wars on which all health and life depend. He got carried away, too, himself in what he was saying, and finally, in complete absence of mind, drank the water in which he had so carefully washed his fruit. I pointed at the glass.

'I am so glad you are still inconsistent!' I said.

'Why, what have I done?'

'Only drunk the water in which you washed your cherries. You've got the microbes safe now, if ever there were any.'

He laughed.

'Well, well,' he said, 'that was indeed careless. I suppose I ought to take an emetic, but even then there is no certainty of complete expulsion. Let us hope for the best.'

I left soon after lunch, and, as I had work in town, did not expect to see Porter again for a week or two. As a matter of fact, I never saw him again, for within a fortnight he was dead of cholera.

PHILIP'S SAFETY RAZOR

UP to the time of Philip's obsession there cannot have been in all the world a happier couple than he and his wife. As everybody knows, the ecstasy of life has its home in the imagination, and Philip Partington and Phoebe lived almost exclusively in those realms which were illumined by the light that never was on sea or land. I do not absolutely affirm that sea and land would have been the better for that light; all that I insist on was that the Partington effulgence certainly never fell there. It was a remunerative light also, and out of the proceeds they bought a quantity of false Elizabethan furniture and a motor-car.

Phoebe and her husband lived in an opulent and lurid present, just as remote from contemporary life as most people know it, as were the 'spacious days' that had left their spurious traces on the Elizabethan dining room.

They were the most industrious of artists, and often had as many as three *feuilletons* running simultaneously in provincial newspapers, and the manner of their activity was this. Every morning directly after breakfast Philip sat in the dining-room, and until one o'clock proceeded to turn into narrative the very complete and articulated skeleton of the tale which Phoebe manufactured in the drawing-room. The imaginative gift was hers; there was not a situation in the world which she could not contemplate unwinking, and these she passed on to her husband, whose power of putting them into narrative was as unrivalled as his wife's in conceiving them.

Picture him, then, with his plump, amiable face bent over Phoebe's imaginings, a perennial pipe in his mouth, and, invariably, two or three little tufts of cotton-wool stuck on to his cheek or chin, on places where he had cut himself shaving that morning. Occasionally, but very rarely, he had to go into the drawing-room to ask the elucidation of some situation—how, for instance, was Algernon Montmorency to leap lightly out of the window, and so regain his motor car, when Phoebe had laid the scene in the top room of the moated tower in Eagle Castle? But Phoebe could always suggest a remedy, and ten minutes afterwards Algernon would be thundering along the road with the lurid Semitic money-lender in close pursuit. But for that and the periodical lighting of his pipe, Philip would not pause for a second till the morning's work was over.

After lunch they drove in the motor car, returning for tea, and from tea till dinner they read over aloud and discussed their morning's work. In this way Philip made acquaintance with the subject-matter he would be employed on next morning, and Phoebe learnt how that which she had written yesterday had turned out.

After dinner they played patience, went early to bed, and awoke with an unquenchable zest for the labour of another day.

It is impossible to figure a happier or more harmonious existence. In imagination they roamed over the entire world, without the expense or inconvenience of foreign travel; their spirits ranged all the gamut of human emotion, and whatever adversities the Algernon and Eva of the moment went through, their creators and interpreters knew in their heart of hearts that all was going to end well, for otherwise they would speedily have lost their pinnacled eminence as writers of serial stories in the daily press.

It is true that sometimes some ill-mannered reader would write to the newspaper office to point out that St Peter's, Rome, did not stand on a 'commanding eminence', or ask more information about the 'glittering spires' on the Acropolis at Athens, or demur to the 'pellucid waters of the Nile in flood, as it rolled down in blue cataracts studded with milk-white foam' . . . But otherwise their life flowed on in an unbroken succession of literary triumph and domestic happiness.

Then suddenly, out of the blue the curtain was rung up on this psychological tragedy, for Philip, by some species of spiritual infection from his wife, began to develop an imagination. It did not at first threaten to attack what Phoebe in a Gallic moment had once called their *vie intérieure*, by which she meant their literary labours, but was directly concerned only with the present of a safety razor which she had made him on his birthday, in order to save cotton wool and his life-blood.

This safety razor consisted of a neat little sort of rake in which razor blades were fitted. Each of these when blunted by use was to be thrown away, and a fresh one inserted, and that morning Philip, finding that his blade had begun to lose its edge, tossed it lightly and airily out of his dressing room window, from which it fell into a herbaceous border which ran along the house. The new blade gave the utmost satisfaction, and precisely at nine-thirty he lit his first pipe and began work on Phoebe's scenario for the day. The dining-room was just below his dressing-room, and at that moment there came a rustle from the herbaceous bed, and Phoebe's adorable Persian cat leapt on to the window-sill from outside, and proceeded to make his toilet in the warm May sunshine. And at that precise and fatal moment Philip Partington's imagination began to work. It stirred within him like the first faint pang of a toothache.

For some quarter of an hour he refused to recognize its existence, and proceeded to clothe in suitable language the flight of the current Eva up the frozen Thames in an ice-ship. Not knowing exactly what an ice-ship was, and being aware that his readers would be similarly ignorant, he evolved a beautiful one out of his inner consciousness that 'skimmed along' on a single runner like a skate.

Suddenly he laid down his pen. His imagination was beginning to hurt him. It would be a terrible thing if Phoebe's cat, while he prowled through the herbaceous bed, stepped on the blade of the safety razor. Blunt though it was for shaving purposes, it would easily inflict a cruel wound on Tommy's paw. When his work was done, he must really hunt for the blade, and bestow it in some safer place.

He took up his pen again, and wrote, 'Ever faster through the deepening winter twilight sped the ice-ship, and Eva controlling the tiller in her long taper fingers watched the dusky banks fly past her. "Oh God," she murmured, "grant that I may be in time!" The woods of Richmond. . . .'

The cat had finished his toilet, and jumped down again into the herbaceous bed. Philip heard a faint mew, and his awaking imagination told him that Tommy had cut his foot already. With a spasm of remorse, he ran out into the garden and began a frenzied search for the razor-blade which with such culpable carelessness he had thrown away. A quarter of an hour's search was rewarded by its discovery, and as there was no blood on the edge of it, he thankfully assumed that he had not been punished (nor Tommy either) for his thoughtlessness.

He locked it up in the drawer of his knee-hole table, where he kept his will and his passbook and his chequebook, and with a free mind returned to work. But suddenly, and for the first time in their dual and prosperous career as *feuilleton* writers, Philip found himself noticing a certain want of actuality in Phoebe's imaginings. They lacked the bite of such realism as he had found illustrated in the poignancy of his own search for the discarded razor-blade in the herbaceous border.

There was emotion, real human emotion, though only concerned with the paws of a cat, and a razor, whereas Eva's taper fingers on the tiller of this remarkable craft seemed to want the solidity of mortal experience. But it would never do to lose faith in Phoebe's inventions, for it was his faith in them that lent him his unique skill as interpreter and chronicler of them. And, anyhow, the razor-blade was safely inaccessible now to any cat on its pleasure-excursions, and he turned his mind back to the woods of Richmond.

With the unexpectedness of a clock loudly chiming, his imagination began to work again. What if he should suddenly die even as he sat there at his table? Phoebe alone knew where his will was kept, and he saw her, blind with tears, unlocking the drawer and groping with trembling hand among its contents.

Suddenly she would start back with a cry of pain and withdraw her hand, on which the fast-flowing blood denoted that she had severed an artery or two, and would bleed to death in a few seconds, as had happened to a most obnoxious Marquis in the tale, 'Kind hearts are more than coronets.'

Next moment he had unlocked the drawer, and, gingerly holding the dread instrument of Phoebe's death between finger and thumb, looked wildly round for some secure asylum for the hateful thing.

Long he stood there in hesitation; then, mounting a set of 'library steps', deposited it on the top of the tall bookcase which held the file of all the newspapers in which their tales had appeared. Then he set to work again. But half the morning had already gone, and he had scarcely yet made a beginning of the morning's work.

Phoebe was unusually buoyant at lunch-time today, but for once her cheerfulness failed in shedding sunshine on Philip.

'My dear, I have got over such a difficult point!' she said. 'Do you remember how Moses Isaacson got Algernon to sign the paper which acknowledged that he was not Lord St Austell's legitimate son?'

'Yes, yes,' said Philip feverishly, trying to recall the exact happening of those miserable events.

'Well, all that was written in invisible ink, and all he thought he signed was the lease of Eagle Castle. There! And look, here is the first dish of asparagus!'

'And how about the lease?' asked Philip.

'It was written in water-colour ink, and, of course, Isaacson washed it off afterwards.'

'Capital!' said Philip. 'That does the trick!'

There was silence for a minute or two as the novelists ate the fresh asparagus, and then Phoebe said:

'Tomorrow, dear, you will have to come and work with me in the drawing-room. The maids must begin their spring-cleaning—and, indeed, it should have been done a month ago. We will have lunch and dinner in the hall while they do this room, and the day after they will do the drawing-room, and I will do my work with you here.'

Philip's fingers were stealing towards the last stick of asparagus, but at this they were suddenly arrested.

'Ah, spring-cleaning!' he said, with assumed cheerfulness. 'They just dust the books, I suppose, and sweep the floor.'

She laughed.

'Indeed, they do much more than that,' she said. 'Every book is taken out and dusted, they move all the furniture, and clean it all back and front and top and bottom. But you won't know a thing about it, except that our dear Elizabethan dining-room will look so spick and span that Elizabeth herself might have dinner in it. Some day we must do a historical novel, you and I. Think what a setting we have here!'

Though the day was so deliciously warm, it felt rather chilly in the evening, or so Philip thought, and a fire was lit in the drawing-room. Phoebe had a slight headache that evening, and thus it was quite natural that she should go to bed early, leaving her husband sitting up.

As soon as he had heard the door of her bedroom close, he went softly to the dining-room, and, again mounting the library steps, took down the razor-blade from the cache which this morning had seemed so secure, and went back with it into the drawing-room. It would have been terrible if Jane, the housemaid, who always sang at her work, should tomorrow have suddenly interrupted her warblings with a wild scream as she dusted the top of the bookcase.

Perhaps the razor-blade would have embedded itself in her hand; perhaps, even more tragically, her flapping duster would have flicked it into her smiling and songful face, and have buried it deep in her eye or her open mouth. But now this gruesome domestic tragedy had been averted by Philip's ingenious perception of the chilliness of the evening, and with a sigh of relief he dropped the fatal blade into the core of the fire.

He went softly up to bed, feeling very tired after this emotional day. Now that his anxiety was allayed, he would have liked to tell Phoebe how silly he had been, for never before had he had a secret from her. But, then, one of Phoebe's most sacred idols in life was her husband's stern, masculine common-sense that (like Algernon's) was never the prey of foolish fears and unfounded tremors.

He hated the idea of smashing up this cherished image of Phoebe's, and determined to keep his unaccountable failing to himself.

He fell asleep at once, and woke in the grey dawn of the morning to the sound, as it were, of clashing cymbals of terror in his brain. The housemaid would clear up the fireplace in the drawing-room and there among the ashes, like a snake in the grass, would be the keen tooth of the razor-blade.

Perhaps already Philip was too late, and before he could get down a cry of pain would ring through the silent house, betokening that Jane's life blood was already spreading over the new Kidderminster carpet, and he sprang from his bed and with bare feet went hurriedly down to the drawing-room.

Thank God he was in time! A minute afterwards he was on his way up to bed again with the razor-blade still dusty with ashes, but as sharp as ever, in an envelope taken from Phoebe's table. Temporarily, he put it between his mattresses, and, since it was still only half-past four, climbed back into bed and vainly attempted to compose himself to sleep.

Already he was behindhand with work that should have been done yesterday morning, and when today, with the envelope containing the blade in his breast-pocket, he tried to make up for lost time, he only succeeded in losing more of it.

There were other distractions as well, for, owing to the spring-cleaning in progress in the dining room, he sat with Phoebe in the drawing-room, and she, quite recovered from her headache and quite undisturbed by his presence, was reeling off sheet after sheet in her big, firm handwriting of the further trials that awaited Algernon.

Falling more and more behind her, Philip lumbered in her wake, with three-quarters of his mind entirely absorbed in the awful problem regarding the contents of the envelope in his breast pocket.

Suddenly, brighter than the noonday outside, an idea illuminated him, and he got up.

'I shall take ten minutes' stroll,' he said. '*Solvitur ambulando*, you know, and you have given me a difficult chapter to write!'

She recalled herself with an effort to the real world.

'I think I won't come with you, darling,' she said. 'I am afraid of breaking the golden thread, as you once called it. Let me see——' and she grabbed the golden thread again.

At the bottom of the garden ran a swift chalk-stream that had often figured in their joint works, and towards this Philip joyfully hurried. He picked up half a dozen pebbles from the gravel path, put them into the envelope which contained the instrument of death, tucked the flap in, and threw it into the stream. There was a slight splash and he saw the white envelope shiningly sink through the water until it came to rest at the bottom. He returned to Phoebe with the sense that he had awakened from some strangling nightmare.

For a couple of days after that Philip enjoyed the ecstasy which succeeds the removal of some haunting terror. Basking in the sunshine of security, he could look back on the dark clouds through which he had passed, and feel that he had really been encompassed by the fringes of some trouble of the brain which borders on the insanity of a fixed idea.

The simple expedient of throwing the razor-blade into the stream had entirely dispersed those clouds, and till then he had never known the sweetness and sanity of the sun. Then with tropical rapidity the tempest closed in upon him again.

He and Phoebe had driven out in their motor-car one afternoon, and had dismissed it two miles from home in order to have the pleasure of walking back through the flowery lanes. Philip was something of a botanist, and since he was now engaged on the chronicling of the reunion of Eva and Algernon, which unexpectedly took place in a ruined temple near Rome, he wanted to refresh his memory by the sight of the glories of the early English summer, in order to deck the flowery fields in which the ruined temple lay with the utmost possible lavishness of floral tapestry.

They had come near to the stream that flowed by the bottom of the garden, the bank of which was a tangle of flowers.

'Loosestrife, meadow-sweet, marsh-marigold, willow-herb!' said Philip. 'Delicious names, are they not?'

The sound of shrill juvenile voices was heard, and, turning a bend in the lane, they came opposite the pool wherein Philip had thrown the razor-blade. There on the bank were half a dozen small boys in various stages of *déshabillé*, rosy from their bathing.

'Little darlings!' said Phoebe sympathetically. 'What a jolly time they have been having in the water!'

'Willow-herb, marsh-marigold,' murmured Philip mechanically, looking round for the traces of blood on the stream-bank.

He took a firm hold of himself, and managed to walk across the wooden bridge that led to the bottom of the garden with some show of steadiness. But he almost reeled and fell when, looking into the pool, he saw the razor-blade, its envelope having been melted off by the water, shining on the pebbly bottom of the stream like some devastating Rheingold.

He made some lame excuse of studying flowers a little further when they had had tea, and slipped down again to the stream. The boys had gone, and, taking off his shoes and socks and rolling his trousers up to the knee, he waded out over the sharp pebbles to where his doom flickered in the sunshine. With the aid of his stick he propelled it into shallower waters and picked it up. Then, shivering from the brisk water, he returned with it to the house in a state of most miserable dejection.

For the next week Philip carried the razor-blade about with him in a stud box that during the day never left his pocket, and at night reposed under his pillow. He made several attempts to get rid of it in a way that commended itself to his conscience that seethed with scruples and imaginary terrors, burying it once in the garden and at another time throwing it into the dustbin.

But the sight of his terrier digging for a suitable hiding-place for his bone in the potato-patch caused him to disinter it from the first of these hiding-places, and the second entailed a dismal midnight visit to the dustbin when, one evening, Phoebe casually alluded to the dustman's approaching visit.

On another occasion he was fired with the original notion of embedding it in the interstices of the rough bark of the ilex at the end of the garden, well out of reach of curious fingers, and, with the stud-box in his pocket, climbed with infinite difficulty up into its lower branches. But while wedging it into a suitable crevice, the bough on which his weight rested suddenly gave way, and he fell heavily to the ground, while the blade flashed through the air like Excalibur and plunged into a bramble-bush. It was, of course, necessary to get it out, and this prickly business, combined with a sprained ankle, brought him almost aground on the shoals of despair.

He began contemplating enlisting as a private in the British Army, though well over the military age and of obese figure. Perhaps he would find some opportunity in Flanders of throwing it, suitably weighted, into a German trench. Only the thought of Phoebe left alone and making up interminable plots with no one to turn them into narrative for her kept him from this desperate step.

Meantime his work halted and languished, for sleepless nights and nightmare days miserably affected his power of composition, his style, and even such matters as punctuation and spelling.

It was when his condition was at its worst that there gleamed a light through the tunnel of his despair. The Editor of the *Yorkshire Telegraph*, who wanted another story by the Partingtons with the shortest possible delay, wrote to him suggesting that the life in New York would present an admirable setting for it, especially since the United States had come into the war, and offering to pay his passage to that salubrious city if he would favourably consider this proposal.

And all at once Philip remembered having read in some book of physical geography, studied by him in happier, boyish days, that the Atlantic in certain places was not less than seven miles deep.

He read this amiable epistle to his wife.

'Upon my word, it sounds a very good plan,' he said brightly. 'What do you say, Phoebe?'

Phoebe shook her head.

'Do you propose that I should come with you?' she asked. 'What about the submarines?'

The thought of the deep holes in the Atlantic grew ever more rosy to Philip's mind. Even the hideous notion of being torpedoed failed to take the colour out of it.

'My dear, these are days in which a man must not mind taking risks,' he said.

She smiled at him.

'I know your fearless nature, darling,' she said, 'but what is the point of running risks?'

'Local colour. There is a great deal in Mr Etherington's remarks.'

'I don't agree. I should think with our experience we ought to be able to describe New York without going there. We didn't find it necessary to go to Athens, or Khartoum, or Mexico.'

'True,' said he, 'but perhaps my descriptions might have gained in veracity if we had. That was a tiresome letter to the *Yorkshire Telegraph* about the spires on the Acropolis. If we had been there, we should have known there weren't any.'

He felt the stud-box in his pocket for a moment, and his fingers itched to drop it over a ship's side in mid-Atlantic.

'My part of our joint work might gain in true artistic feeling,' he said, 'if I described what I had actually seen. Art holds the mirror up to nature, you know.'

'Yes, darling, but do you think Shakespeare meant that art must hold the mirror up to New York?' asked she. 'I fancy there is very little nature in New York.'

He took a turn or two up and down the room, while the box positively burnt his finger-tips.

'I can't help feeling as I do about it,' he said. 'And, Phoebe, one of our earliest vows to each other was that each of us should respect the other's literary conscience!'

She got up.

'You disarm me, dear,' she said. 'Apply for your passport, and if they give it you, go. I only ask you to respect my feminine weakness and not make me come with you, among all those horrid submarines.'

They sealed their compact with a kiss.

By the time Phoebe had interviewed her cook, her husband had already written his letter applying for his passport, on the grounds of artistic necessity in his profession. She read it through with high approval.

'Very dignified and proper,' she said. 'By the way, dear, there will be no work for us this morning. We are going over the factory for explosives with kind Captain Traill. You and I must observe the processes very carefully as we want all the information we can get for "The Hero of Ypres."'

He jumped up with something of his old alacrity.

'Aha, there speaks your artistic conscience,' he said. 'And don't let me see too many soft glances between you and kind Captain Traill.'

Phoebe looked hugely delighted, and returned the compliment.

'And there are some very pretty girls working there,' she observed slyly.

An hour afterwards they were padding in felt slippers round the room where bombs were packed with a fatal grey treacle, one spoonful of which was sufficient to blow them and the whole building into a million fragments. A new type of bomb was being made there, consisting of a cast-iron shell fitted with a hole through which the grey treacle was poured; an iron stopper with a fuse in it (released by pressing a spring) was then screwed into the hole. There were hundreds of those empty shells, which slid along grooved ways to where the treacle was put into them, and they then were passed on to the work-girls, who fixed their stoppers. It was all soft, silent, deadly work, and Philip recorded a hundred impressions on his retentive memory.

Phoebe and Captain Traill were walking just ahead of him when suddenly a great light broke, so vividly illuminating his brain that he almost thought some terrific explosion seen and not heard had occurred. Stealthily he drew from his pocket the stud-case, stealthily he opened it and took out the razor blade. Then, bending over an empty bomb-case as if to examine it, he dropped the blade into it. It fell inside with a slight clink, which nobody noticed.

A couple of minutes afterwards the bomb-case had passed through the hands of the dispenser of treacle, and had its stopper screwed in.

'And where are all those little surprise packets going?' asked Philip airily.

'To bombing aeroplanes on the West Front,' said kind Captain Traill. 'We're sending off a lot tonight. Perhaps that one——' and he pointed to the identical bomb which Philip had had a hand in filling, 'will make a mess in Mannheim next week.'

'I hope so,' said Philip fervently.

The only thing, now that Philip had disposed of the razor-blade, which clouded his complete content was the fear that his passport would be granted him, and he would have to make a journey to America. Happily, no such unnerving calamity occurred, for a week later he received a polite intimation from the Passport Office that the object for which he wanted to go there did not seem of sufficient importance to warrant the granting of a permit. So, wreathed in smiles, he passed this letter over to Phoebe.

'There's the end of that,' he said.

He took up the morning paper.

'Hullo!' he said. 'Our airmen bombed Mannheim two nights ago, and dropped three tons of high explosives. Well, that is very interesting. Captain Traill said that perhaps some of those bombs we saw being filled would make a mess in Mannheim. I hope it was those actual ones.'

'So do I,' said Phoebe. 'Was there much damage done?'

'The German account says that there was hardly any, but, of course, that is the German account. A few people were wounded and cut by fragments of bombs. Cut!'

He got up, and could hardly refrain from dancing round the table.

'Some deep cuts, I shouldn't wonder,' he said. 'Gashes!'

THE HAPLESS
BACHELORS

IT had been a shattering blow to those two middle-aged bachelors when Mrs Nicholson announced her miserably selfish intention of getting married.

For ten years she had managed their house for them with economy and sublime efficiency, cooking and serving for them their punctual and delicious meals, calling them in the morning, and bringing in their bedroom candles to their sitting-room at night when she thought it was time for them to go to bed, valeting them, telling them to order new clothes when necessary, reminding them when it was seemly to have their hair cut, ordering coal and wine when required, checking their bills and thwarting unscrupulous trades-people, counting the very heads of asparagus in the bed and the strawberries under their nets, and, in a word, taking off their hands the whole burden of those interminable domestic details in which, like midges on a summer evening, life is so annoyingly enveloped.

There was a gardener, whom she controlled (for the rolling of the croquet-lawn, the production of abundant vegetables, the glowing evidence of flower-beds were outside her immediate management), and there was a girl who came in at break of dawn and worked under Mrs Nicholson's sweet supervision till she staggered home at sunset, exhausted in body but wonderfully enlightened in methodic industry. But before she came Mrs Nicholson would have lit the kitchen fire, and laid the grate in the dining-room and sitting-room, if she thought the weather sufficiently inclement to justify such a proceeding, and after the girl had gone she would serve dinner and do some sewing. There was not such a word as 'fatigue' in Mrs Nicholson's dictionary; if there had been she would not have comprehended it.

Six months of intolerable discomfort succeeded for her two unhappy masters, Mr Beaumont and Mr Bradley, and though they were of normal and kindly dispositions, they became frankly misogynistic under the blast of the incompetent females who now witheringly blighted their household.

Instead of one benignant and all-pervasive personality who allowed them to lead smooth and untroubled lives and devote the whole of their leisure to their own hobbies, they were harassed and rendered miserable by three inadequate females—a cook, a parlour-maid, and a housemaid—who between them did with notable inefficiency what Mrs Nicholson had done with the speed and precision of a planet moving in its splendid orbit.

The cook could not get on without a charwoman, the parlour-maid without two evenings out a week, and the housemaid without breaking Beaumont's choicest pieces of Salopian china and knocking down Bradley's setting-boards on which were impaled his latest captures of butterflies and moths. The machinery of life creaked and groaned under these nincompoops, and however often they were replaced, the change never produced any better results.

Neither Beaumont nor Bradley had a moment's leisure. The cook's melancholy face would succeed her mournful rap on the door, and she would announce that the fish had not come, or, having come, had gone bad; the parlour-maid would rustle in to ask what wine she was to open for dinner; the housemaid would want some new dusters.

Buttons came off their coats and were not replaced, lamps smoked, things went wrong with the supply of hot water, letters were not posted, devitalized syphons refused to give up their contents, keys were lost, books were put upside-down into their shelves, and a gramophone wheezed nightly in the kitchen.

The general disintegration, too, was not confined to the house, for, with the departure of Mrs Nicholson, a similar deterioration set in in the garden.

The croquet-lawn, that smooth sheet of green velvet, began to grow bumpy; the asparagus-bed provided only a week's refreshment;

the early potatoes were a complete failure, and the roses had blight. None of these things had been in Mrs Nicholson's direct province, but her effulgent example had shone on the garden no less than the house, her supreme efficiency had infected that once productive acre.

Worst of all, most ruinous to content and leisure, were endless minute complaints. 'It's not what I've been accustomed to, sir,' was the burden of them, delivered in various cold and acid voices. As for the household bills, they rose like snow-drifts in some bitterly inclement winter.

Toothache had taken Beaumont to London one bright June morning (no tooth of his had ever ached in the reign of Mrs Nicholson), and while waiting to enter the torture-chamber, his attention was attracted by an advertisement in a magazine which was headed, 'No more Blood or Cinders for Dinner'.

This extolled the merits of an electric oven, which roasted or grilled or fried whatever was submitted to its ministrations. No trained cook—here Beaumont quite forgot about his tooth—was necessary: the householder merely ascertained by consultation of a small hand-book supplied gratis with the oven, how long a sole or a mackerel or a chop required for its perfect presentation at table, turned the dial on the oven for the required number of minutes, and could rest assured that a delicious dish awaited him.

Should he desire a joint prepared with the same perfection, he had but to weigh his joint and allow for its cooking its weight in pounds multiplied with the number of minutes required for each pound.

Milk, water, and vegetables could be boiled by the application of a simple formula; in fact, there was no act of cooking which the 'Ichabod-oven' did not accomplish with invariable nicety.

At the expiration of the minutes indicated on the dial (brass with enamelled face), the heat was automatically cut off.

There was no more stoking of the kitchen fire, no more waste of coal; the 'Ichabod-oven' had but to be connected by a cord to a plug in the wall, it cooked, and it cut off its own current. Should the cooking be done and the current cut off before the viands were needed, there was nothing easier, when the last guest had arrived, than to switch on the Ichabod again, and heat up in a couple of minutes the cooling delicacies.

A picture of a smiling hostess welcoming a guest who was half an hour late for dinner, owing to a breakdown on the Tube, endorsed and crowned the merits of the Ichabod. 'That makes no inconvenience in our house,' she benignantly observed. 'George will turn on the current to the Ichabod-oven, and in two minutes you will have your dinner as hot as hot.'

Beaumont always propitiated his dentist with pleasant conversation when his mouth was not full of wads, and he selected this subject for his spasmodic remarks. Between whirlings of the electric wheel and manual jabs and stabs, his torturer upheld from personal experience all that the advertisement had claimed for the Ichabod-oven, and told Beaumont of a dozen labour-saving apparatuses of the same kind, which could be seen at Messrs Milliken's furnishing establishment in Bond Street.

'Electricity,' he observed, as he applied the whizzing wheel to Beaumont's tooth, 'takes the place of manual energy. One moment: a little wider, please! There!'

Beaumont had no time that day to test the merits of these admirable contrivances, and arrived home to find Bradley in the most morose spirits. The housemaid in a fervour of dusting had whisked her napkin over one of his collecting-boxes, plucking the wings and antennae off two Purple Emperors and six Clouded Yellows. Even as he recounted his irremediable disaster, Beaumont, who had opened a cabinet containing his choicest pieces of Salopian china, gave a shrill squeal of dismay.

'Who has broken my toy Salopian milk-jug?' he asked.

'Oh, Jane,' said Bradley. 'It flew out of her hand, she told me, as she was dusting it.'

'But she had no business to dust it,' cried Beaumont.

'I know she hadn't, nor my Purple Emperors. We must give her notice. It's your turn.'

'I can hunt for ten years without finding another toy Salopian milk-jug,' moaned Beaumont. 'Yet even the worst servants are getting scarce.'

At dinner the two carved in turn, week by week about. As Bradley plunged a blunt knife into the leg of lamb, a torrent of blood poured out, as if he had beheaded a vampire.

'Totally uneatable,' he said. 'We must cut some slices and have them grilled.'

The slices were cut and sent back to the kitchen. After a long delay there returned some thin crackling cinders.

'First blood, then cinders,' observed Beaumont, and into his mind there came the remembrance of the advertisement he had seen. The loss of his toy Salopian milk-jug had put it and his tooth out of his head. Now as he crashed into a cinder they both returned to him.

Though Beaumont had been to town that day, an hour's discussion with his friend decided them! He and Bradley went up again next morning, and after a succulent lunch prepared exclusively by the Ichabod, spent the entire afternoon at Messrs Milliken's.

Long before tea-time—with hot buttered scones from the Ichabod—they had completely made up their minds, and having ascertained that every labour-saving device in stock could be installed in their house in three weeks, they dismissed the entire household with a month's wages instead of a month's warning, and moved across to the admirable hotel, where, in comfort, they could superintend the refitting of their home.

Electricity, that stern but obedient force, was to take the place of a staff of greedy, incompetent females; in the garden alone was male labour to be retained, since weeding and the care of vegetables were not yet within the scope of electrical energy, but within doors the two intended to take on themselves the entire labour—a matter, it appeared, of some half-hour a day—of the house.

A crippled soldier with the resounding name of Fotheringay would make beds and empty slops, which was sordid employment; but when once these were done, there was nothing which could not be effected swiftly and without loss of dignity.

It was positively exhilarating to hold a muddy shoe to the bright, whirling electric brush; five seconds sufficed for the complete removal of mud, five more for its superb polishing.

In the kitchen stood the resplendent Ichabod, which cooked dinner for ten—should that be necessary. On the kitchen-table, over which no longer querulous females stuffed themselves four times a day with expensive food, was the Universal Washer-up. Into its convenient racks you placed the sullied plates and dishes and teacups and knives and forks, filled it with hot water from the Ichabod, and connected it with a plug in the wall. The Universal Washer-up then took on a majestic rocking movement, and in five minutes the implements of feeding were spotlessly clean and the hot water incredibly dirty. A tap let out the foul fluid, and a little more exercise on the part of the Universal Washer-up sufficed to dry its contents.

Where the kitchen-range had been, there was now a stove which heated the boiler to such an extent that oceans of scalding water rumbled in the bath-room and squirted from the hot-water taps that steamed in every bedroom.

Central heating had been installed all over the house, and the same supply of water rushed madly through miles of radiators. There was a machine like a small grass-cutter so light that it could be taken up in one hand, which had only to be connected with a plug in the wall to cause it to give forth thirsty, sucking sounds. When pushed over a carpet, it absorbed into a small chamber every particle of dust that had lurked there. This provender was invaluable for garden-beds.

A whirling brush of soft feathers dusted everything that the thirsty grass-cutter could not reach and pleasantly ventilated the rooms. Then, as there would be no permanent resident in the kitchen, it was undesirable that tradesmen should enter, and so in the middle of the kitchen-door was inserted a commodious safe. The tradesman—butcher or baker or grocer—had but to press a conspicuous button, and the door of the safe flew open. He then inserted in it such goods as had been ordered by telephone, together with the bill, closed the door of the safe again, and went away, conscious that if he had not placed there what was ordered and what was detailed on his invoice, detection would inevitably await him.

A complicated sewing-machine undertook all hemming and stitching, and by a special adjustment put on buttons. Clothes, after the ministration of the grass-cutter, were pressed by the turn of a screw; and all cotton, woollen, or linen garments were washed by the Universal Washer-up in its off-moments. But the starching-machine was not yet quite perfect, and the Messrs Milliken did not recommend it.

For a couple of days after this inclusive instalment an intelligent young man remained on the premises and cooked dinners and supplied hot water, and washed plates and pressed clothes, and dusted with a perfection that Mrs Nicholson might have envied.

One morning Bradley tore his trousers quite badly over barbed wire in the pursuit of a Humming-bird Hawk moth, and the young man in two minutes, with the aid of the sewing-machine, repaired

the rent in a perfectly amazing manner. The entire housework
seemed to occupy him but a couple of hours in the day, and he said
that he could run half a dozen houses single-handed without turning
a hair.

With a final injunction not to employ the full current for the lift
which conveyed the hot fruits of the Ichabod to the dining-room, he
got on to his motor-bicycle and disappeared with the speed of the
Vanishing Lady.

Beaumont had the first day of housework (for they had settled not
to share these duties, but each of them to perform all of them on
alternate days) and produced an excellent supply of hot water for the
morning baths. Owing to some oversight with regard to the Ichabod,
the eggs for breakfast were boiled as if for canaries, but the bacon
was crisp and good.

Fotheringay's notion of making beds was sketchy, for he merely
replaced the bed-clothes, and he emptied the slops into the bath, but
promised to do better in future.

About the middle of the morning loud and agonized yells from
outside the house caused Beaumont to conclude that something was
wrong somewhere, but he did not think it worthwhile to interrupt his
dusting to investigate the matter, for these screams probably came
from the house of a noisy Irish family a few doors off, where the
children were always quarrelling.

An hour or so later, having occasion to visit the kitchen, he was
astonished to observe the lower part of a human arm in the safe
inserted into the back door, and found that it had slammed to with
incredible violence as the grocer was in the act of putting a pound of
sugar into it. The poor man was faint with pain when he was
released, and Beaumont, lucidly explaining the mechanism of it, had
the end of his nose severely slapped as the door unexpectedly flew
back again.

The grocer, humbly but firmly, announced his intention of, for the
future, leaving his parcels outside that deadly trap. This rank
cowardice had something in its favour, for he pointed out that fresh
butter, if left as long as his arm had been left in the safe, would not
be materially improved by contact with raw meat, whiting, candles,
and possibly onions.

The rest of the day passed in a blaze of efficient mechanism. No
gramophone had interrupted the studious quiet, no servants had

protested against things they were not accustomed to, while the financial gain seemed to Beaumont, as he added up the bills for goods supplied, to be most encouraging.

'Electricity doesn't spend the day in guzzling in the kitchen,' he finally observed. 'Electricity doesn't interrupt one every minute with complaints. What peace to know that there isn't a woman in the house!'

Bradley breathed a sigh of sympathetic relief.

'I entirely agree,' he said, 'as far as female servants are concerned. Women may be, in fact, they are, ornamental little things; they have a brightness, a charm about them. For the purposes of the propagation of the species, they seem to be essential.'

'Doubtless; but you and I do not propose to propagate the species. And I was talking about their domestic efficiency.'

'There was Mrs Nicholson,' began Bradley.

'But did you ever, even in Mrs Nicholson's reign, spend a more tranquil day than to-day?' asked Beaumont.

'No, I can't say I did. The eggs at breakfast——'

'Which you ate with considerable relish in the salad at dinner,' interrupted his friend.

'Quite true. I withdraw the eggs. The imprisoned grocer, however. I think, perhaps, we should be wise not to continue using that safe. It would, by the way, make a magnificent breeding-box for my caterpillars.'

A faint humming sound had been audible during this conversation, and Beaumont visited the kitchen to see whether any electric energy had been left running. He found that he had forgotten to turn off the apparatus that cleaned boots, and a pair of his own which he had left in contact with the revolving brushes were almost on fire with the friction. The post had just come in, and he opened a letter from his widowed sister, Mrs Glover, who usually spent the month of August with them, proposing her annual visit. She would bring down with her, if convenient, her new French maid, Hortense, who ran her house for her with exquisite finish, and was really more of a friend and companion than a servant. A most superior woman, and pretty enough, as she observed, to capture even Mr Bradley's hitherto impregnable heart. This last sentence made her brother frown; Mabel had set her becoming cap at Bradley for years.

'Mabel suggests herself for early in August,' he said on his return.

'She proposes to bring a marvellous maid, a sort of Mrs Nicholson. We'll teach these women a lesson as to how to run a house!'

'And no message for me?' asked Bradley.

'Some silly joke about the good-looks of her maid. Mabel is rather vulgar sometimes. Bless me! it's nearly eleven. How quickly time goes when one is peaceful and tranquil! I'll put the burglar-alarm shutters up.'

He got one rather severe shock from making the electric connection before he let go of the bell.

Bradley's day began inauspiciously. He must have made some mistake in the taps that controlled the hot and cold water, for every drop of water in the house speedily rose to boiling-point. It was impossible to shave or to wash or to have a bath, for scalding fluid poured out of both taps alike, wherever situated. Boiling water thrummed and chuckled everywhere, the whole central-heating apparatus was in full operation, and this, added to the heat of the morning, made the house absolutely untenable, and unwashed and unshaved and streaming with perspiration, they breakfasted in the garden.

The Ichabod, on the other hand, however much Bradley turned taps and adjusted dials, developed no heat whatever; and though they could have plenty of tea by filling the teapot at any tap in the house, nothing could be cooked. Eggs, however, could be boiled in the bath, and before long the slight initial error about the cold-water supply was remedied.

Bradley had another smart shock when he took the burglar-alarm shutter off the pantry window, but consolation for his tingling arm was administered by the discovery that he had merely forgotten to connect the Ichabod up with the supply of electricity.

Beaumont offered to help with the dusting, and unfortunately brought the whisking feather-brush too near a shelf of Salopian china. Plates and teapots rose in the air like a covey of partridges. Meantime, Bradley, by a too-energetic application of the grass-cutter machine, caused it to suck up not only the dust from a valuable Persian rug, but a strip of nap. A lane was cut in it like the path of a tornado.

He then transferred his caterpillars into the discarded safe in the back-door, but as he forgot to close the door, they crawled all over the kitchen. He recaptured most of them, and began work on the Ichabod for lunch.

There is no need to dwell on the horrors of the next three weeks. From native inefficiency combined with steadily accelerating nervousness, if one of two taps had to be turned, Beaumont and Bradley alike turned the wrong one. The Universal Washer-up was altogether forgotten one night after dinner, and continued to rock backwards and forwards till all the plates and glasses and dishes came out of their racks and formed a heap at the bottom. A soup-ladle which should never have been put there at all dealt stunning blows at the defenceless crockery and speedily reduced it all to shards.

In the night the burglar-alarm continued to ring at intervals until dawn. But with the pathetic obstinacy of men they continued unbroken in their determination to show Mrs Glover and the fascinating Hortense how vastly more efficient was the male sex in all domestic matters, and, the weather continuing very hot, they laid in, at staggering expense, quantities of cold delicacies which did not require cooking.

Hortense, on arrival, modestly established herself in the kitchen, and professed a delightful willingness to help in the housework; so Beaumont explained to her at considerable length the merits of the Ichabod and left the preparation of dinner in her hands.

She appeared to be most intelligent, and promised to do her best. It was only for tonight, so Beaumont assured her, that they would trouble her, but he had not seen his sister for so long, and he had many things to talk over with her. He had not, however, any immediate opportunity of doing so, since Mabel and Bradley had already gone for a walk, and did not return till close on dinner-time.

Hortense sent them up, serving it herself, a most delicious repast, and four perfect courses inspired a spirit of boastful insincerity on the part of Beaumont.

'You see, my dear Mabel,' he said, 'what can be done by two bachelors. You have dined, as you know, not badly——'

'Perfectly, exquisitely!' said Mabel. 'Too good for words!'

'Well, cooking is child's play if you have an Ichabod. That's the sort of dinner Bradley and I get every night. All the wages and bother of a cook are saved. I can cook just such a dinner in half an hour. No knowledge of anything is required. Bradley will cook for us tomorrow.'

'Ah, but how clever of you, Mr Bradley,' said she. 'And you will be housemaid too? Men are wonderful! I always said so.'

Hortense, one way and another during the next week, seemed to have taken charge of the house. When Bradley's day came he explained to her the virtues of the whirling feather-brush and the dust-sucker, and warned her against the employment of the Universal Washer-up unless strict supervision was given it. While Beaumont merely endorsed all his friend had said.

As a matter of fact, they were both revelling in the smoothness and perfection which recalled, and indeed exceeded, the reign of Mrs Nicholson; both looked forward with something remarkably resembling dread to the fatal day when Mabel Glover would depart and take her Hortense with her. Dire misgivings, never communicated and internally suppressed, occasionally seized them when they came into a room and found Hortense sweeping a carpet without the aid of the grass-cutter, or dusting shelves without the whirling circle of feathers, but as a rule all housemaid's work was over before they came downstairs, and Fotheringay somehow had melted away, and was no more seen.

All this was observed and forgotten, and Beaumont, with slightly more difficulty, managed to forget the fact that on visiting the kitchen one day when the Ichabod should have been at the height of its activities, he found that his culinary expert stood cold and with open doors. Hortense explained that just for tonight she had not used it.

The smooth, delicious days slipped by; neither Beaumont nor Bradley now made the smallest pretence of going into the kitchen at all, and week by week, though they were all pampered with toothsome viands, the house-books showed a splendid economy.

Bradley, who, on the whole, was the greedier and the more luxuriously-minded of the two bachelors, viewed with the greater despair the resumption of their usual bachelor establishment. The prospect of that appeared to him increasingly insupportable, and increasingly pleasant seemed the companionship of Mabel Glover, who helped him to set his moths with wonderful delicacy of touch, and was never tired of procuring fodder for his imprisoned caterpillars.

Essentially, he was hooked already; practically, two days before Mabel's proposed departure, he proposed to her and was instantly landed. But what, so thought the impassioned swain, would happen if Beaumont, goaded by ignoble revenge, proposed to Hortense?

No such heart-breaking calamity occurred. The only thing that Hortense insisted on was that electricity should only be used for bells and illuminations, and that she should have no cage for caterpillars in the door of *her* kitchen.

DICKY'S PAIN

DICKY PEPYS, up to the age of fifty, had lived an extremely happy, selfish, and innocent life. He had lost two tiresome parents while he was yet in his 'teens, and at the age of twenty-one had come into a very ample fortune, unencumbered with the wretched hardships of rank and of land-owning, except for that portion of the earth's surface upon which stood his charming house in Berkeley Square. His substantial inheritance had put him out of reach of the Bankruptcy Court, his prudence and innate respectability were shining lights to a naughty world, his teeth rendered him absolutely uninteresting to any dentist, and all the ostriches in Africa might have envied him his superb digestion. He had thwarted all the designs of impecunious Countesses, who longed to saddle him with their daughters; he never provoked the retribution that falls on those who have strained their hearts by excessive athleticism, or on those who by their sedentary habits make an enemy of their livers. He always went to Scotland in the summer, when he made charming sketches of the moors over which more barbarous folk shot stags, and in the winter he went to Monte Carlo and sat in the sunshine while others lost fortunes at the gambling-tables.

His mind was almost—not quite—as healthy as his body, but it had one little kink in it. As he sat in the sun or sketched in Scotland, he often permitted it to imagine lurid situations. He wondered what he would do if he lost all his money, he wondered what he would do if by some appalling mischance he fell in love, and he wondered what he would do if he ever felt unwell.

His house in Berkeley Square was a model of convenience, for his father had been a notable hypochondriac, and Dicky lived in his suite of rooms on the second floor. The bath-chair in which his father was wheeled to the passenger-lift still stood at the end of the passage, and every morning Dicky made his ablutions in a bathroom, which was fitted with squirts and douches of hot and cold, and an electric-chair which, when you sat in it and turned a switch, proceeded to jog about and encourage the internal organs. There were wash-leather pads attached to steel arms, which you could lay on the place that hurt and then give it massage, and there was a pair of scissors with immensely long handles with which you could cut your toe-nails, if lumbago made it inconvenient to bend the back. All these cunning contrivances Dicky had kept, partly out of filial piety, but partly, also, out of a curious internal interest in them. Heaven forbid that they should ever be needed by him, but where was the use of scrapping them?

In the middle of July he always left London and went down to his bungalow at Littlehampton, where he indulged in croquet, sea-bathing, and golf on the ladies' links, till it was time to go to Scotland to sketch, and on this particular afternoon in mid-July he had tripped up the stairs (not using the lift, of course) to inspect what his valet had put out for him to take to Littlehampton. He would want flannel shirts for playing croquet, and thin shirts for playing golf, and a bathing-costume, and some sandals for walking over the stones, and some packs of patience cards, as well as the ordinary packs for bridge, and some menu-cards for dinner and . . . But his admirable valet assured him that the bungalow was plentifully stocked with all these things, and even as he spoke, Dicky suddenly felt a slight pain in the middle of his front, just below his chest. It lasted only a minute or two and then he felt quite comfortable again. Probably it was some minute indigestion, and he thought that he would not have vermouth before dinner. It might be due to vermouth, which, he had been told, was not quite wholesome.

He was going to motor down to Littlehampton, and he had three guests arriving there that evening from London. Naturally, he had not proposed driving them down, for they would have luggage which he could not take, and they would no doubt feel safer if they went down with it by that excellent train that arrived in good time for dinner. Besides, two of them were females, and if he took them with him, he would have to sit on the front seat, which made him feel rather giddy, or go on the box, and put his valet inside. In fact, he would be much more comfortable alone, and they would feel much safer with their luggage in the train with them. Among his guests was Lady Earlswood, who, after trying to marry him herself, was now trying to marry him to her daughter, who was another of his guests. But Lady Earlswood (even if he had meant to marry anybody) was too old, and her daughter was too young. The third guest was Claude Bingham, a gay young man of about forty, who would be delighted to flirt with either of them.

It was a superb evening, hot and windless, and after dinner they all sat in the garden of the bungalow. Dicky had abstained from vermouth, and felt particularly well: he even announced his intention of having a dip before breakfast next morning.

'Better not, Dicky,' said Lady Earlswood; 'people like you and me, who are getting on in life, must be careful.'

Dicky thought this rather malicious: in any case, Lady Earlswood had got much further on in life than he: five years further at least. But he was constitutionally good-natured, and bore no resentment.

'Fancy you talking of getting on in life, Bridget!' he said. 'I never heard such nonsense. I know how you danced till morning the other day at the Carews'.'

'Yes, and suffered for it afterwards,' said Bridget. 'My dear, I felt like nothing at all next day. Did I, Lucy?'

Lucy was an anaemic young woman, quite unlike her florid mamma.

'You weren't so bad as I was, Mamma,' she said. 'I wanted you to come home earlier. I haven't felt well ever since.'

To Dicky's immense surprise, Claude Bingham echoed these depressing sentiments.

'I can't sit up late now,' he said, 'and be alive next day. Port has begun to poison me, too. I took some tonight, Dicky, for your port is irresistible. But I shall have heartburn tomorrow.'

Dicky felt it incumbent on him to strike a more cheerful and robust note.

'Heartburn?' he asked. 'Never heard of it. What and where is it?'

Rather to his vexation, Claude pointed to the precise spot where he had experienced that little pain this afternoon.

'Indigestion of a sort,' he said. 'I have to be careful of it.'

Dicky got up.

'Come, this will never do,' he said. 'We shall all get pains if we think about them. Let's play Animal Grab. I shall be a nightingale. Jug, jug.'

Animal Grab produced a healthier outlook, and after an excellent night Dicky trotted off across the garden in his pretty dressing-gown with his bathing costume below it for his morning dip. He splashed about in the tepid shallow water for two or three minutes and enjoyed it so much that he determined to have a 'real' bathe, including swimming, before lunch. It was wonderful, he thought, as he trotted back across the garden, to be so juvenile at the age of fifty, and as that vainglorious thought passed through his brain, he felt another little pain just where he had felt it yesterday. It couldn't be vermouth; was it possibly sea-bathing? Perhaps he would not have a real bathe today. On the other hand, the pain yesterday couldn't have been sea-bathing, since he had not bathed for a whole year.

His party hailed him as a marvellous athlete when they assembled for breakfast, which was pleasant, but after a game of croquet he slipped away under the plea of writing letters, and telephoned to a doctor in the town of whom he had heard well and went to see him. Dr Bannister was most reassuring, and recommended strict moderation (or better, abstention) in alcohol, and a glass of very hot water to be sipped half an hour before meals. A few bending and swinging movements of the trunk on getting up and just before going to bed would be a healthful practice, and he might also paint the troublesome spot with iodine. There was nothing to be anxious about. Sea-bathing? Well, an intermission of sea-bathing for a day or two would do no harm.

Dicky resolved to get rid of this little pain, and in order to make a firm stand against it, he cut off wine altogether, scalded his mouth with boiling water three times a day and used iodine regularly. But three days afterwards he experienced it again for a few minutes, and he wasn't sure whether there was not another pain a little higher up in the region of the heart. In consequence, when at the end of the week, Claude, who had been drinking quantities of port and suffering much from heartburn, concluded his visit, Dicky drove him up to London and went to see Sir Francis Tollington of Harley Street. He shook his head over the hot-water treatment, and recommended abstention from starchy foods, like bread and potatoes, and assured Dicky that he need not worry.

'And how about the iodine?' asked Dicky.

'My private opinion is,' said Sir Francis, 'that you might as well put it on the door-mat. But it can't hurt you unless it takes the skin off.'

'And sea-bathing and croquet and golf?' he asked.

'All excellent in moderation,' said the doctor. 'But cut off starchy foods.'

Dicky enjoyed pastry and adored potatoes, but he heroically followed this advice and remodelled his diet. His garden just now was profuse in delicious fruits, notably figs, and he made up for the new restrictions by a copious indulgence in them. It seemed to suit him admirably, but one afternoon, just as he made a beautiful drive from the 18th tee of the ladies' links, he had a return of the pain. It was not the least severe, but all the pleasure from his beautiful drive evaporated like breath on a frosty morning. As he approached the last green, he saw standing outside the club-house the burly form of his old friend Dr Samuel Janitor, the great specialist on diseases of the nervous system, who shook hands warmly.

Dicky was delighted to see him: the pain was still perceptible, and really these dietists didn't seem to do any good.

'Well, this is delightful,' he said. 'You must come to dine with me tonight, Sammy. We shall be quite alone, and will have a good game of chess.'

'Rather. And how are you, Dicky?'

'Pretty well,' said he.

Dicky had no compunction about consulting a doctor when he was on holiday, and Dr Janitor heard all about it before he was allowed to play chess. He was most convincing, pointing out that in middle life nervous force began to fail, and, if the body was to remain in perfect health, it must be supplemented. He believed that Dicky's pain was certainly connected with the solar plexus, an important junction or nest of nerves which was situated just where the trouble was. Electricity could supply the evident deficiency of nervous force, and the treatment was quite simple and could be practised at home. The apparatus was rather expensive, but once purchased the cells lasted a long time, and could be cheaply recharged. Half an hour, morning and evening, was all the time required, but regularity was essential.

Dr Janitor made some more percussions in the region of the solar plexus.

'You should be careful about your diet,' he said. 'You oughtn't to touch fruit for the present; be very sparing in the use of tea and coffee, and cut off tobacco altogether. There is nothing that makes such a drain on the nervous system.'

This was an unpleasant hearing, for Dicky liked his cigar even more than he liked his potatoes. But he was determined to get his solar plexus in condition again, and he put back into the box the fresh cigar he had just taken.

'That's right,' said Dr Janitor. 'You'll soon be all right if you'll just discipline yourself. You'll find that a little hot lemon-juice and water takes away any craving for tobacco. Now for our game.'

Though so many of the normal pleasures of life were cut off, Dicky was so keen on regaining his usually perfect health that he gladly gave them up. The treatments he had been recommended were very varied, but any one of them might be successful and so he practised them all. The battery arrived in a day or two, and what with applications of it morning and evening, and hot water before meals, and iodine paintings, and sips of lemon-juice, he was getting busy. He felt very well, however, and it was a cruel disappointment that on the very day that he was returning to London, meaning to go up to Scotland the day after, the wretched pain came again. It was very slight, scarcely perceptible indeed, but he realized that he was not cured yet. Perhaps it was something much more serious than either Dr Bannister, Sir Francis Tollington, or Dr Janitor had suspected, and he telegraphed to Sir Augustus Boughton and made an appointment for next morning at his house in Harley Street. He felt that he could not enjoy Scotland at all unless his mind was relieved of the grim fear which now cast a shadow over it, and at eleven o'clock next morning he presented himself, rather shakily, at Sir Augustus Boughton's, which happened to be exactly opposite Sir Francis Tollington's. He hoped the latter would not be looking out of his window.

Sir Augustus was gaunt and cadaverous. He asked Dicky an enormous quantity of questions and took note of his answers in a huge ledger. At the end he read them over in dead silence, and Dicky got more frightened every moment.

'It isn't malignant disease, is it?' he asked in a faltering voice.

Sir Augustus turned on him his mournful gaze.

'I can't tell you until I've thoroughly examined you,' he said.

The examination took place, and again Sir Augustus wrote in his ledger. Then he wheeled round to the trembling Dicky.

'There are no symptoms of malignant disease at present,' he said. 'Your trouble without doubt arises from acidity. It is some gouty or rheumatic affection, and I strongly recommend a course of baths at Slipton Spa. Three weeks at Slipton with a suitable diet ought to be very beneficial.'

'But must I go there instead of Scotland?' asked Dicky. 'I was going to Scotland tonight.'

'I am advising you to do what I should do myself,' said Sir Augustus austerely.

Dicky did not hesitate.

'Very well, I'll go to Slipton,' he said.

'I think you are wise,' said the doctor. 'Now what treatment have you been having hitherto?'

Dicky closed his eyes, and ticked off the items on his fingers.

'I use the Fergus electric battery for half an hour morning and evening,' he said, 'and put on an application of iodine every other day. I sip half a pint of very hot water before each meal, and hot lemon-juice and water when I want to smoke. I never touch fruit, bread (except toast), pastry or potatoes or wine or tobacco or tea or coffee, and every morning and evening after the electricity I make swaying and bending movements.'

'All very sensible,' said Sir Augustus, 'though I don't think a glass of sound Burgundy with your dinner would hurt you, and an occasional cigarette might do no harm. I should also allow you an orange in the middle of the morning. But I regard it as essential that you should completely abstain from butcher's meat. I don't mind your having a little boiled—not grilled—fish at lunch, and a chicken's wing at dinner, but no beef, mutton, veal, or pork. I will write fully to my colleague, Doctor Paley, in whose hands you may place yourself with the utmost confidence. I expect you will receive great benefit from your stay there.'

Dicky arrived next day at Slipton, and sent a note to Dr Paley to ask for a consultation. The doctor, who had that morning received a long letter from Sir Augustus, was most cheerful and encouraging and mapped out his cure.

'Better have your bath in the morning,' he said, 'at half-past ten or eleven, and after it to take a rest for an hour and a half. A real rest, mind, not just sitting in an armchair, but on your sofa or your bed. That will bring you to lunch-time: rest after lunch, and then I want you to go for a gentle walk of forty minutes. You will find the public gardens very pleasant, and after your walk you can go for a run in your motor, well wrapped up. After tea——'

'I never have tea,' said Dicky.

'So much the better. At half-past five then, I should like you to be massaged, and rest afterwards till dinner. Sir Augustus has told me he has already spoken to you about diet, and I may say I entirely agree with his views. You ought to go to bed not later than half-past ten.'

'And when shall I come to you again?' asked Dicky.

'In three days' time, please. Let me say at 4.30 on Thursday. By the way, don't feel disappointed if after a week or a couple of weeks you feel no improvement, but perhaps rather the reverse. The treatment is often lowering for the time, but you may confidently expect great improvement afterwards.'

Dicky was now fairly launched on the sea of hypochondria. The original visitation in the region of the solar plexus had not occurred since the morning he left Littlehampton, but a whole host of other symptoms had flown to join it, as if it had been a decoy-duck. He distinctly felt a twinge in his left knee as he walked one day in the public gardens, and one morning, as he came from his bath there was a slight singing in his ears, and once after rising swiftly from a very low chair, he felt a momentary giddiness. He had no appetite for his lunch one day, but the next he had a disquieting ravenousness, and that night he slept ill. These symptoms were so frequent and varied that, despairing of remembering them all, he jotted them down in order to keep Dr Paley fully informed. . . . In the hotel he became immensely popular, for he took elderly ladies for drives in his motor, and taught them new Patiences in the evening, and listened with sympathetic interest to their symptoms, occasionally ejaculating, 'Yes: I had that yesterday morning, but it passed.' Then, when it was his turn, he described a sense of numbness in his right wrist, and Mrs Moule had it too sometimes. But bathing it in hot water, and then a little gentle massage with the finger-tips, generally relieved it.

Dicky felt that for all these fifty years he had missed his vocation, which was clearly that of an invalid. He had often before, when perfectly well, been rather at a loss what to do and think about, but now there was never a dull moment. Owing to the prohibition about meat, he had been forced for sheer pangs of hunger to eat some starchy foods, but when, after the blissful consumption of three large potatoes at lunch, he sneezed many times when he was being massaged, it was so thrilling to wonder whether he was not somehow throwing off the ill effects of the potatoes. Mrs Moule thought it very likely, and produced the parallel instance of that tiresome cough—Mr Pepys probably had noticed it—which she had last night when they were playing Patience. Well, she was sure it was the hot buttered scones she had for tea.

Dicky's three weeks were drawing to an end, and, in spite of these interests, he certainly felt very much run down. He had had no return (as far as he could remember) of his original symptoms, but there had been so many others. Dr Paley, in his final interview with him, produced the list of them.

'There's clearly some poisoning going on in your system, Mr Pepys,' he said, 'and with the most careful observation, I have been able to find no adequate reason for it. I think it must arise from your teeth.'

'But look at them,' said Dicky gaily, opening his mouth very wide. He always liked his teeth being looked at, so white and regular and complete were they. No dentist had ever been able to insert his finest probe into any cavity.

Dr Paley looked.

'I don't assert that all your sufferings come from your teeth,' he said, 'but I can find no other possible cause for them. If it's not teeth, your case baffles me. Have them all out. They've probably been poisoning you for years.'

'Good gracious!' said Dicky, very much depressed at the thought. But he was now so devoted a disciple of hygiene, that he never dreamed of rebelling against this drastic prescription, and promised to go to see his dentist at once.

Dicky had himself comfortably packed up in his motor for his three-hours' drive to London, and settled down to a good solid and uninterrupted meditation about his health. He was quite prepared to lose every tooth in his head, if he could only regain his health, and would willingly have had a couple of toes amputated as well, if he could find a physician who recommended it, but by degrees it struck him that if his teeth were to blame, his solar plexus and his nervous system were innocent and he needn't think any more about uric acid. There was no need to go to Slipton again or use his electric battery or drink hot water or abstain from fruit, pastry, meat, tea, coffee, tobacco and alcohol.

'I'll see if I can't get Claude to dine with me tonight,' he thought to himself, 'and we'll have the best and biggest dinner I can think of, with wine and liqueurs and coffee and cigars, and peaches. If my teeth are going away, I'll give them a good send-off. Nothing that I eat and drink will hurt me, if it's only teeth. And tomorrow, I'll have them out. I suppose that will mean slops for a fortnight, but I'll stoke up first.'

Dicky, half starved by his abstention, ate a joyful and gorgeous dinner. He and Claude went to the movies afterwards and then had some supper. Next morning after an excellent night Dicky woke feeling extraordinarily well and happy. He had some early morning tea (he had not tasted it for many weeks and found it delicious), disregarded his electric battery and his iodine bottle, and felt himself ready for breakfast. Then he had a cigar, and thought he would telephone to his dentist presently.

Some grouse had arrived from Scotland for him, so he collected a few stranded September friends, and arranged a small dinner-party for the evening: it was therefore impossible to part with his teeth that afternoon. One of them urged him to spend the week-end in the country, and so he put off till next week.

A few days more, if he had been poisoned for years, could make no difference.

Dicky was sitting on the Terrace at Monte Carlo one brilliant morning during the next winter talking to Lady Earlswood. She had been complimenting him on his appearance of amazing health.

'Yes, I'm very well again,' said he, 'though I had some horrid weeks of illness in the summer. Continuous attacks of pain. But I took it in hand seriously, and had electric treatment and iodine and very strict dieting, and cut off wine and tobacco and tea and coffee and spent three weeks at Slipton. Brine baths and massage.'

'How horrid for you,' said she. 'And attacks of pain? Where?'

Dicky frowned slightly as he lit his cigar.

'Dear me, where was it?' he said. 'I really can't remember. But my memory has been getting very bad lately. I must have some course of mental training.'

SOCIETY
STORIES

THE BRIDGE FIEND

SCENE: *Drawing-room at Lady Swindon's country house.*
TIME: *Today, in the month of October, an hour after dinner.*

Enter from the conservatory LADY WITHAM (*the Bridge Fiend*),
followed by MRS SPENCER.

THE BRIDGE FIEND: At last, at last! Really, I think we have stolen
away rather successfully. But I thought Lord Swindon was
literally *never* going to finish the story he was telling me; in fact, he
never did, because he went fast asleep in the middle of it. Did you
ever notice, dear, that he goes on smoking with perfect correctness
when he is asleep, but always lets his cigar go out when he is awake?
Where are the two men?

MRS SPENCER: They are following. You know Lady Swindon has
no idea of the duty of a hostess, which is to leave one completely
alone. She is always exerting herself to be agreeable, whereas the
truest hospitality consists in taking away from your guests the
impression that they are in somebody else's house.

THE BRIDGE FIEND: Then we have assisted her in her duties,
which is the part of the guest. Really, I think that to join in a flow of
polite conversation is the most fatiguing thing that can happen to
one—so enfeebling to the intellect, too, especially sitting in the open
air. I have been flowing, leaking rather, for the last hour and a half.
Well, a rubber will restore us, I hope. To swear at one's partner is
most exhilarating after one has been polite for so long. One will have
plenty of opportunity with Jack Pirbright and Willie Comber.
Neither of them can play at all. But I could see no one else except

Mr Arbuthnot, and I simply can't afford to play any more with him: he always wins. Ah, here they are!

Enter JACK PIRBRIGHT *and* WILLIE COMBER.

Let us cut immediately, because somebody else is sure to want to come and cut in. People are so selfish (*she cuts*). Queen!

WILLIE: Yes, that was pretty smart of me, wasn't it, Jack? You know when I'm really awake I'm far from being fast asleep; and under those conditions, if I really give my mind to it——

THE BRIDGE FIEND: You don't say so! How very droll of you, Mr Comber! Please cut.

MRS SPENCER (*cutting*): Eight!

JACK PIRBRIGHT (*cutting*): Eight!

WILLIE COMBER: I think you'll find this is a king (*cuts*). Oh no! It appears to be a two.

THE BRIDGE FIEND: Eights cut again. Do be quick!

MRS SPENCER (*cutting*): King!

JACK (*cutting*): King!

THE BRIDGE FIEND: Oh, sit down, sit down anywhere, just as we are! I hear a step. Give me the cards. You and I, Mr Pirbright (*deals rapidly*).

Enter MR ARBUTHNOT.

THE BRIDGE FIEND: What a pity, Mr Arbuthnot! You are just too late. We were looking for you everywhere to make up a fourth, and had to sit down without you. Yes, and now we're in the middle of our game. Most exciting.

WILLIE: I say, it was my deal, wasn't it? I cut the two.

THE BRIDGE FIEND: Sh—sh! (*Hastily conceals scoring card* as MR ARBUTHNOT *approaches the table.*) Do come back again soon, Mr Arbuthnot, and cut in. (*Exit* MR ARBUTHNOT.) Mr Comber, how rash of you! Do you claim the deal?

WILLIE: Yes, certainly. I cut the lowest. You and I, Mrs Spencer. Sixpences, I suppose?

JACK: You know I'm a rotten player, Lady Witham.

THE BRIDGE FIEND: I do, to my cost. I remember playing with you before, and a variety of painful circumstances are imprinted on my mind. And please don't make no-trumps again on three knaves and a six, relying on the probable strength in my hand, particularly if I have already passed to you.

WILLIE: I say, how these cards do stick together! It's like dealing bits of court-plaster. I think I've misdealt. I say, it's deuced lucky the deal doesn't pass when you misdeal, isn't it? Oh no, I haven't misdealt. I always wonder why the deal doesn't pass if one misdeals. I can't see why, can you, Lady Witham?

THE BRIDGE FIEND: Please don't talk, Mr Comber. If anyone *utters* during the deal, my attention completely wanders from the game, and I never know what is happening. The other night only, when we were playing at Julia's, there was a man in the corner who was saying *the* most interesting things about—well, about him and her and that sort of thing, and with one ear I couldn't help trying to catch what he was saying, and I had to attend with the other eye. The consequence was——

WILLIE: Well, it's diamonds, then. What were you saying, Lady Witham?

THE BRIDGE FIEND: I double.

WILLIE: Lord, how rich! I redouble.

THE BRIDGE FIEND: Are you content, partner?

JACK: No, not in the very slightest. I never saw such a vile——

THE BRIDGE FIEND (*resignedly*): My dear Mr Pirbright, there is no earthly reason to tell them what a bad hand you've got. They will see it in time: there is no hurry. That information is probably worth a couple of tricks to them.

JACK: Well, that's more than we are ever likely to get out of it. And I should like to see anybody else.

MRS SPENCER (*acidly*): Shall we proceed?

THE BRIDGE FIEND: I was waiting to do so, dear, till we had finished talking. A small club.

MRS SPENCER (*laying down her hand*): The seven of diamonds— my only one—may eventually prove to be of value if you nurse it very carefully, Mr Comber. Otherwise, I see nothing of interest except six little hearts to the nine, which cannot possibly be brought in. And diamonds are worth twenty-four, I believe.

[*A pause.*

THE BRIDGE FIEND: Have you no diamonds, partner?

JACK: None of any kind whatever. I never had. I didn't deal: it's not my fault.

THE BRIDGE FIEND: Did I suggest that?

[*A pause.*

WILLIE: Yes, that makes us the odd, and if anyone can beat the king of diamonds, the ace being out, now's their time. Two odd, forty-eight—thanks for the double, Lady Witham—and four honours in my own hand, which makes forty-eight above. Rather fat on the whole. I like a fat game——

THE BRIDGE FIEND: Would you kindly cut, Mr Comber?

WILLIE: Rather! Particularly when I win it. When the other side wins I think I prefer it lean, like Jack Spratt's wife. By Dr Watts, isn't it? Old Jimmy Methuselah held four honours in his hand the other day when I was playing with him; and he made rather a smart remark. He said he considered it right honourable, like a Privy Councillor. Made me laugh a lot, that! Good old Jimmy!

MRS SPENCER: What very extraordinary friends you seem to have, Mr Comber!

WILLIE: Oh, he isn't a friend exactly. He's a sort of acquaintance——

THE BRIDGE FIEND: If I may get in a word edgeways, I should merely like to remark that there will be no trumps of any description.

WILLIE: I'll tell you about him afterwards, Mrs Spencer.

MRS SPENCER: That will be thrilling. May I play?

WILLIE: Rather! This hand would poison a rattlesnake.

[MRS SPENCER *plays.*

THE BRIDGE FIEND: Please lay down your hand, partner. What a terrific exposure! Dear me, I always believed in chiromancy, that you could tell a man's character from his hands. And do I understand that you've got no heart?

JACK: None whatever. I offer you my hand without my heart.

WILLIE: I say, that's rather like 'My true love hath my heart', isn't it?

THE BRIDGE FIEND: Yes, quite amazingly like. There are four clubs, are there not, headed by the knave? That appears to be your strongest suit.

JACK: Yes, I'm afraid I've got a hand like a foot.

THE BRIDGE FIEND: A club-foot?

WILLIE: That's rather good. I'll try to remember that. Old Jimmy——

JACK: Another heart led, partner. What do you wish me to play?

THE BRIDGE FIEND: Kindly let me think a moment (*pause*). Whichever you like. I don't care in the slightest. No, certainly not that. There!

[*Dead silence for two minutes.*

WILLIE: And the best spade, I suppose, in your hand, Lady Witham? Oh, you chucked it! Then that's the odd to us, and aces easy. Ripping!

THE BRIDGE FIEND: Remarkably ripping.

[MRS SPENCER *deals.*

WILLIE: Oh yes, I was just telling you about old Jimmy Methuselah. He isn't really a friend of mine, only he's always asking me to lunch at the Saltspoon, or some such club, and I never go. I seem to remind him of lunch, somehow, and I can't think why. I suppose some day I shall ask him to dine at the Sugartongs, and he will accept. That will spoil it all.

THE BRIDGE FIEND: Yes, pray be careful not to rub the bloom off so romantic and wonderful a friendship.

WILLIE: I'll bear it in mind. What did you say trumps were, partner?

MRS SPENCER (*wearily*): I asked you to name them some minutes ago.

WILLIE (*with alacrity*): Oh, did you? I never heard you. Well, I name spades.

THE BRIDGE FIEND: Doubled!

MRS SPENCER: Redoubled!

JACK: I'm on in this piece. I make them sixteen. Everybody satisfied?

THE BRIDGE FIEND (*ominously*): We shall be able to tell soon. Two tricks take us out, partner.

[*Dead silence till the end of the hand.*

WILLIE: So we get the odd. Makes us twenty-eight. And honours? Simple.

THE BRIDGE FIEND: May I ask what you redoubled on, partner?

JACK (*with abandoned cheerfulness*): Oh, general debility all round. I am a bit *debile* tonight.

THE BRIDGE FIEND: And is it your habit to trump your partner's best card?

JACK (*with the same cheerfulness*): Yes, I often do it. It makes it safer, if you understand.

THE BRIDGE FIEND: Perfectly! Deal, please. The cards are usually cut first. They are one game and twenty-eight.

JACK: And what are we?

THE BRIDGE FIEND: Thirty for aces the deal before last, also twelve for your chicane in diamonds. All above.

JACK (*gaily*): Oh, then I've contributed something by—by my absence. That's a beginning, isn't it? We'll play with hearts.

WILLIE: Shall I, Mrs Spencer?

MRS SPENCER: Please.

THE BRIDGE FIEND (*laying down her hand*): You will observe, partner, that I have four aces in my own hand. One hundred neatly gone to the dogs.

JACK: Oh, no! they'll all come in useful. Each takes a trick, and each trick is worth eight. That's thirty-two out of the kennel (*pause*). Yes, this time we'll play that pink ace. Dear, dear, it gets trumped! Who would have thought it? I wonder if I ought to have played it first round?

THE BRIDGE FIEND (*coldly*): If your object was to take tricks, there was not the slightest cause for wondering. It could not have been otherwise.

[*Silence.*

JACK: And in romps the last trump, like the Day of Judgement. Three odd, twenty-four, and all the honours.

THE BRIDGE FIEND (*rapidly*): Apart from the question of that deplorable ace, partner, it did not seem to strike you that if you had finessed the queen, you could have led through the clubs and placed the lead here, discarding your only remaining diamond on my ace, roughing the spade next time, and forcing the second hand to trump, thus placing the lead on your left, and getting your hearts led up to instead of through.

JACK: I beg your pardon, Lady Witham—would you say it again?

THE BRIDGE FIEND (*rather more rapidly*): Isn't it perfectly clear that there must have been only one spade on your left, and that if, instead of leading as you did, you had opened clubs from here, you could have finessed your queen, then led through the strength in clubs, since the ace was staring you in the face opposite? Then you had only one diamond left, which you could have discarded on my ace of spades, and have roughed the spades next time, which your leading them originally prevented, since Mr Comber had only one, and could subsequently over-rough you. Then he must have put on his only remaining trump, and your queen and ten in trumps would both have made. Isn't that clear?

JACK: Yes, perfectly, now you explain it. Amazingly stupid of me!

WILLIE: Cut, Jack, will you? (*Aside*): I say, I should take a nip of something, old chap, or all that will lie pretty heavy.

JACK (*aside*): No fear; I didn't listen. When will this be over?

WILLIE: Well, we're twenty-eight. I rather think I'll—I'll pass.

THE BRIDGE FIEND: Hearts, did you say?

WILLIE: No, I leave it.

THE BRIDGE FIEND: I understood you to say 'hearts'. Didn't you think he said 'hearts', partner?

JACK: No, Lady Witham; I thought he said 'pass'.

THE BRIDGE FIEND: I don't know what the rule is in such a case.

MRS SPENCER: The rule is, dear, that if the dealer passes, his partner makes trumps. I make diamonds.

[*A thundery silence, but the electrical tension is relieved when* THE BRIDGE FIEND *and her partner have secured the odd.*]

THE BRIDGE FIEND: Yes, Mr Pirbright, you played that better.

JACK: Ah, I remembered what you made so clear to me about finessing something and putting the lead in the other place.

THE BRIDGE FIEND (*graciously*): So glad! That is game all, then. Cut, please, Willie (*he cuts*).

THE BRIDGE FIEND: I leave it.

JACK (*hastily*): No trumps. I've got——

THE BRIDGE FIEND (*with a shriek*): Silence!

MRS SPENCER: How you startled me, dear! May I?

WILLIE: Yes, by all manner of means, as my Billy Henley always says.

MRS SPENCER: How very entertaining of him! A small heart.

JACK (*laying down his hand*): There, Lady Witham! That's a rise in my character, isn't it? Four aces and two kings. What a picture!

THE BRIDGE FIEND: Yes; the makings of a hand.

WILLIE: I say, that looks rather on the fat side of lean, doesn't it, partner?

[A long silence.

THE BRIDGE FIEND: There! Four odd. Puts us out. Rather ticklish work, was it not, at one point, Mr Pirbright? I was afraid that the nine and the five were both on the left, so that if I discarded the eight, and chanced there being no more diamonds on the right, the remaining——

MRS SPENCER: Yes, dear, quite so, but your last card is a spade.

THE BRIDGE FIEND (*brazenly*): Yes, the best.

MRS SPENCER: Oh, quite the best—only you revoked.

THE BRIDGE FIEND: Revoked? What are you talking about, dear? You see, in that case, Mr Pirbright, we should have lost both, and only got two odd, which would not have been sufficient. I had nothing in my hand, as you saw.

MRS SPENCER: You had a spade, dear Violet, which escaped your notice. I will look at the second—no, the third trick, please.

THE BRIDGE FIEND: Pray do.

[High wrangling ensues; the revoke is irrevocably established.

MRS SPENCER: So we'll take the value of three tricks, please, which is thirty-six, and wins us the rubber. The hundred will exactly cancel with your hundred aces.

JACK: Dear, dear, what an unfortunate oversight! I'm so glad it wasn't I.

[A somewhat painful silence as the score is added up.

THE BRIDGE FIEND: What a small rubber! A few shillings only, is it not?

WILLIE: Yes, that's all.

THE BRIDGE FIEND: Mr Pirbright, I——

JACK: Oh, pray don't apologize! It may happen to anyone. It was——I'm sure if you were going to revoke, you revoked in the best possible manner.

THE BRIDGE FIEND: Ah, the revoke! No, I was just going to explain to you about those clubs. It was really quite an interesting point. You see, if Mrs Spencer had held both, we should only have got two odd.

JACK (*the worm turning*): Yes, dear Lady Witham, I am sure you are right. On the other hand, if you had not revoked——

THE BRIDGE FIEND: Let us play another. Cut in quick, for fear anybody else should come. As we sit. Cut, please, Mr Comber.

[*She begins to deal. Enter* MR ARBUTHNOT.

MR ARBUTHNOT: Still at it?

THE BRIDGE FIEND: Dear Mr Arbuthnot, we were looking for you. Yes, hearts are trumps. We've just begun another. How very unfortunate! Do stop and cut in with us at the end of this. (*Enter two other guests.*) Or perhaps you could make up another table. Yes, hearts. No doubling? What masterly inactivity on the part of our opponents! Spade led? Yes. You don't seem to cultivate hearts, partner. Really, I never saw such a hand! I think I should not play at all if I couldn't hold better hands than that. Yes, the little one.

WILLIE: Before which I place the queen.

THE BRIDGE FIEND: Dear Mr Comber, please don't talk so much; it takes off one's attention so. Yes, I put the king on to that. She falls, does she not, poor dear? The ace of trumps from my hand (*pause*), followed by the king (*pause*). You see, partner, if I had played a little one, I should not have captured the queen. One has to go for the big game occasionally.

MRS SPENCER (*bitterly*): Yes, dear, but as you hold all the rest of the trumps in your own hand, the risk was not so great.

THE BRIDGE FIEND: Please, dear Isabel, *after* the game is over. I cannot play if people talk. That is better. So we get all the rest.

THE DRAWING-ROOM
BUREAU

MRS AUDLEY had left London the week before the war began, but she lost little time in putting her charming house in Curzon Street in commission again, when it was clear to her that London and not the country was to be the home of civilized people this autumn. She herself was intensely civilized; indeed, she had almost civilized herself away, so to speak, and really only existed in the midst of other people where she could hear what was being said, and say it herself. It was this latter instinct that led her to aspire to the position of Drawing-Room Bureau, and spend her days in retailing news connected with the war, and not officially published.

Decidedly she had gifts which fitted her for her self-created post. Physically, she had height and distinction; materially, she had money, which she liked spending in the entertainment of other people; and psychically she had a delightful manner when talking to anyone, which conveyed the impression that this particular opportunity for conversation with this particular individual was to her the crown and fulfilment of her existence. In her vague violet eyes there dwelt (how it had got there I cannot imagine) a look of earnest and limpid sincerity, and, like all those who have a very clear consciousness of their mission (hers being Drawing-Room Bureau), she was almost completely devoid of humour. And though one did not feel any overpowering confidence in her communiqués, it was impossible not to be slightly flattered when she beckoned with her serious eyes and made a place on the sofa close beside her.

'Perhaps I oughtn't to tell you,' she would say, 'but I know it won't
go any further. It is very doubtful if we shall send an Expeditionary
Force to France at all. I was in Whitehall this morning, though I
mustn't say exactly where, nor who was my informant, though I can't
help your guessing. Well, the matter is still under discussion, I can
tell you that, and today the discussion was hot. Mind, I don't say
there is any serious discussion between, well, between A and B, but
what somebody feels is that if we guarantee, as we are doing, the
safety of the French coast, that is a good deal. Hush! There's Mr
Armine coming; we will talk about something else. But drop in
tomorrow—let me see, not at five, but at twenty minutes to five—
when we shall be uninterrupted, and I may be able to tell you more,
though I can't promise. Good evening, Mr Armine! How unkind you
are! You haven't let me have a word with you yet.'

Now it was impossible not to feel, so confidential was the manner
in which Mrs Audley distinguished you thus, that you had not been
admitted into the secret councils of the War Office or Admiralty
(whichever it was that had been visited by her this morning), and
that she had taken you, as by some discreet and delicious back-door,
into the very presence of A and B. You might have a fleeting
scepticism about it all when, subsequently, you found out that on the
very evening when she had told you that the Expeditionary Force
was not yet finally decided on, it had already arrived in France, or
feel less personally flattered when it appeared that she had told Mr
Armine exactly what she had told you. But such impressions soon
faded, and when next Mrs Audley confided something that must on
no account go any further until Tuesday at the earliest, you listened
with undiminished avidity, and almost marked off the days as they
passed, like a schoolboy with his calendar that records the approach
of the holidays. She served her information up so appetizingly, like
some consummate chef, and it tasted so good that it was out of the
question to consider seriously what it was made of.

All through the autumn her standing as Drawing-Room Bureau
grew steadily in importance. Despite occasional lapses, she was very
alert to find out what was likely to happen, and the very fact that she
never directly told you who her informant was added a piquancy to
her secrets. She was Delphically mysterious with regard to the two
hundred thousand Cossacks who were supposed to have passed
through England on the way to the French battle-line; but her

reticence here, you felt, was of the nature of a thin layer of ice over impenetrable depths and boiling springs.

'You will be right,' she said, 'not to believe half you hear about trains passing through Willesden and Swindon with all the blinds drawn down. Mind, I don't say that there were not such trains, but, as far as I know—and my information is, I think, the latest that is authentic—there was no movement of troops through Swindon, anyhow. As to the transports to and from Archangel about which you asked me just now, it is quite true that the *Oceanic* was up there, and that she has gone down. And I can tell you that Archangel is still clear of ice, which is unusual, and that at present it is being kept clear. I was at the Russian—well, perhaps I had better not say exactly where—I was at a certain house last night, and heard some very strange things. Don't you wish I was like the majority of my sex, and didn't mind whether I was pledged to secrecy or not? But at present I can't tell you more. Ah, here is Lady Weyburne; let us make room for her. Dear Claire, come and talk to us. We are longing to know what Italy is going to do.'

Now this, without doubt, was meant to be a disarming remark, but it unhappily failed in its intention, and only made Lady Weyburne buckle on her armour instead of taking it off. She was, in fact, beginning to be a serious rival to Mrs Audley in this Bureau business, and was getting on so well that probably Mrs Audley wanted to amalgamate. But Lady Weyburne had no idea of doing so; when she had rivals she did not want to propitiate them, but to pommel them. She ran her establishment on very different lines from Mrs Audley's discreet and personal methods. She made no quiet confidences, but told stories that she had heard or invented to rooms full of people at the top of her shrill and raucous voice. Already the rivalry had led to tiresome situations, as, for instance, when she announced to the entire dinner table that Sir John French had been in London, at the very moment when Mrs Audley was imparting that fact to the man on her right (in Italian, so that the servants could not understand what she was saying) as the most precious of all secrets that must on no account be known till Wednesday evening.

Tonight it was evident that Lady Weyburne, instead of being disarmed by this request for information, flung it back, so to speak—a gauntlet of challenge.

'Dear Madge,' she said, 'fancy you of all people asking poor me for information. Why, I am told on all sides that the only person who knows anything about the war is you. Indeed, I was on the point of asking you if it is really true that immense quantities of macaroni and Capri wine have been shipped to Marseilles for the consumption of the Italian troops that arrived there on Sunday afternoon?'

Mrs Audley's serious mind scented war.

'No, dear, I have heard nothing about that,' she said, 'but then I hear so little. How interesting it must be to be told these wonderful things! Capri wine and macaroni! Dear me! I remember, darling, how you thrilled me with delicious stories of the Russian troops passing through Swindon, and knocking the snow off their boots. I wonder what has happened to them all. Have you heard anything further?'

Lady Weyburne laughed; she laughed as if she was thoroughly amused.

'My dear, what can you be dreaming of?' she said. 'I never believed in the Russian troops at all, in spite of all you said about them.'

'No?' said Mrs Audley dreamily.

'Of course not. There never were any. I can't think why your friends at the Embassy made such a mystery of it, if you really understood them correctly. No, I won't sit down; I am just off to the Foreign Office. Ah, perhaps I had better not have said that! Forget it, won't you?'

Now here was a declaration of war, a challenge as to Drawing-Room Bureau, as direct as words could possibly make it, for the last sentences were a pure parody of Mrs Audley's unmistakable style, and from that moment the Bureau war may be said to have definitely begun. Instantly both sides, already completely mobilized, violently attacked and counter-attacked. It was enough for Lady Weyburne to learn that Mrs Audley had made the faintest suggestion about the attitude of Romania, to cause her to quote her cousin at the Foreign Office, with or without his authority, in support of her contradiction of whatever Mrs Audley had said. This was a real, genuine cousin (though whether he was strictly responsible for all the war-news which Lady Weyburne fathered on him is another question), and all the information which Mrs Audley hinted that she had received in letters from her husband who undoubtedly was at the front could not quite hold their own against these cousinly pronouncements, especially since everybody knew that letters from the front were heavily censored and probably did not contain so much direct news as Mrs Audley hinted at. On the other hand, the latter had a really magnificent innings about Christmas when Major Audley got four days' leave; after his departure the town rang with stories of remarkable strategy, which Mrs Audley told everybody, with strict injunctions that they must never be repeated at all, not even after next Tuesday.

On the whole, then, the honours of the campaign were for many weeks about equally divided, and neither side could claim an unchallenged supremacy, nor completely round the other up. In point of industry and imagination the two were very equally matched, though in different styles, and while Mrs Audley was content to confide to you that a wonderful new type of gun, about which the utmost secrecy must be observed, had been landed (though she mustn't say where) in France, Lady Weyburne boldly announced its calibre and range, confusing inches with millimetres and miles with kilometres, without the smallest embarrassment or hesitation. Though this wealth of amazing detail carried you off your feet for the moment, Lady Weyburne lacked the impressiveness of Mrs Audley and her subtly dropped hints that the Ambassadors of the Allied Powers were in the habit of dropping in to tea and pouring out their hopes and fears. Mrs Audley, in fact, produced atmosphere; her Turneresque mists and gleams were full of suggestion, while Lady Weyburne reminded you more of some Dutch painter in whose landscape you could count the leaves of the trees and the number of hairs on the cow's tail. Some minds preferred one, some the other, but all alike waited with breathless interest to see the outcome of this brave rivalry. It came early in March, when Lady Weyburne won all down the line by a stroke so Machiavellian that she almost ought to have forfeited all claims to be considered a civilized being. It certainly was war, but morally it could not be called magnificent. The fiendish manner of it was in this wise.

In spite of the deadliness of the struggle the two were otherwise on perfectly friendly terms. The area of conflict in which neither gave nor asked for quarter was strictly circumscribed to the question of supremacy in the dissemination of Admiralty and War Office news not officially promulgated. Where that was concerned there was no act of hostility which they would not cheerfully perpetrate on each other, but apart from that their relations were completely amicable. In the ordinary course of life, then, Mrs Audley, by appointment, went to tea one afternoon with Claire Weyburne, and on arrival was told that she had been unexpectedly detained, but would be in presently, and hoped that Mrs Audley would wait.

Mrs Audley did so, and, as was natural, wandered round her friend's newly decorated room, appraising and approving. A Chippendale bureau stood open, with some charming examples of Copenhagen china on the top, and on the flap a few papers and a blotting-book. Having admired the china, her eye fell (it was no more than that) on the papers, and she found it impossible not to observe that one had the insignia of the Foreign Office stamped on it. It followed that she could hardly help seeing what was written on it. It ran thus:

DEAR CLAIRE,

I send you a communication which will interest you. The message was taken from a German wireless. Keep it to yourself, won't you, as there are reasons why it should not be made public at present, if indeed ever.

Your affectionate cousin,
S. BURGIN.

Mrs Audley was not probably much more dishonourable than most of us. But she was at this moment transplanted to the dire field of battle, and her ordinary human scruples gave but one thin, sad cry and expired. She saw that a further sheet was attached to this covering letter, and she took the two up, not observing that a pin that lightly held them together slipped out, and, turning the covering letter back, she read:

German wireless reports that one of the enemy's cruisers in the Pacific has taken Ipecacuanha. The Admiralty admits this, but doubts the ability of the enemy to retain it. The internal disturbances that will almost certainly result——

She broke off suddenly, hearing a step on the stairs, and hastily replacing the papers moved swiftly away from the bureau. When Lady Weyburne entered she was absorbed in the Japanese silk curtains which hung at quite the other end of the room.

'Darling, so sorry not to have been in,' said Lady Weyburne, not looking anywhere near the Chippendale bureau, 'but I simply couldn't help it. Have you been here long?'

'Just a couple of minutes.'

Lady Weyburne, still with eyes averted, rang the bell for tea.

'And we meet again this evening,' she said, 'for Daisy Johnston told me you were dining with her. But that shan't prevent us having a good talk now. Tell me what you think of my room. You have such taste.'

This was not in the least ironical, and during the course of a chattering hour Mrs Audley made several excellent suggestions. She even had nerve enough to look closely at the Chippendale bureau, just as if she had not seen it at all before her hostess arrived. And she did not glance even for a fraction of a second at the letters and papers that lay there. . . .

But when she had gone Lady Weyburne looked rather closely at them. She observed that a pin which held two papers lightly together had fallen out. Then, after sudden and unexplained laughter, she sat down quietly to think, with a broad grin on her acute and pleasant face.

That evening, accordingly, they met at dinner, and it might have been noticed that Lady Weyburne did not, as was her usual custom, sit down to play Bridge afterwards, but engaged in trifling conversation near the fireplace. Mrs Audley never played, and, as was usual with her, sat and talked in a low voice to a man for whom she made room on the sofa. Her voice was quite inaudible to Lady Weyburne, but that she expected. The man in question was Dick Ransom, who was keenly interested in the fortunes of the Drawing-Room Bureau war.

'I am afraid that things are not going very well in the Pacific,' said Mrs Audley to him. 'Do you remember that German cruiser which escaped and made its way up some river in East Africa? We thought we had prevented her getting out again, but I know that I was never quite easy about it. I'm afraid she must have escaped, for I have heard a disquieting item of news. If you will promise not to let it go further, I can tell you. She has taken Ipecacuanha—indeed, the Admiralty admits it—though they say that owing to the internal disturbances that will probably result, the enemy will not be able to retain it. Very likely you will never see the taking of it officially announced——'

Suddenly Dick Ransom gave a great shout of laughter, and on the moment Lady Weyburne rose, interrupting her trivial conversation, and came towards them. Without a shadow of doubt she had been waiting just for that, for Dick's uproarious laugh.

'Ah, do tell me what the joke is,' she said. 'I long to hear something amusing in these dull days. Madge, darling, what have you been telling Mr Ransom?'

Dick laughed again, slapping his great thigh.

'Best thing out,' he said. 'A German cruiser has taken Ipecacuanha, and the Admiralty think they mayn't be able to retain it. Ha! You'll never beat that, Mrs Audley!'

Lady Weyburne looked enquiringly at her rival, who rose suddenly, with a certain hunted and dismayed expression in her violet eyes.

'Ah, that silly joke of my cousin's,' she said. She paused a moment, and her laughter shrilled high above Dick's.

'But you told it Dick as serious war-news?' she asked. 'Dear Madge, you should get somebody to censor your war-news for you, in case it happens to be a joke. And who could have told you this, I wonder?'

But she was very busy for the next day or two in telling absolutely everybody exactly what she wondered.

MUSIC

WHEN Elizabeth Nutter came to London with a view to perching herself on the very topmost bough of the pleasant tree of Society which grows by the Thames, she possessed the usual equipment of such amiable invaders. She had an amusing face, gowns from Paris, pearls from the Orient, an iron, inflexible will, any amount of money, a variety of complexion to suit all lights, and no husband. She had given a considerable amount of thought to this last negative equipment before she left the Argentine Republic, but had decided against providing herself with one. In some ways, of course, a husband might be of use, for to have a tame, quiet man in the background, ready to be moved, as required, to the middle distance or the foreground, was a sort of guarantee of respectability, but she had come to the conclusion that a guarantee of respectability was not in London worth the paper on which it was written. In New York, she understood, such guarantees were occasionally demanded for inspection like passports, but in London a really convincing guarantee of unrespectability was more likely to be an asset, for London notoriously welcomed and made much of many of those who were socially expelled, as being undesirable natives, from New York. Besides, a husband might grow skittish or want to have his own way, instead of remaining the gentle domestic animal that husbands should be, and which her late husband undoubtedly was. Also if she brought a husband to London she could not buy a new one there, without a great deal of trouble and publicity in disposing of the first. She did not object to publicity—indeed, she did not really care about anything else—but it had to be the right kind of publicity, not the sort of inquest which was held in the English divorce courts. In any case, it would be well to make herself at home in London before she chose this permanent addition to her life's appurtenances, otherwise he might not suit with her social colour scheme, which at present was undetermined, and it was well to see what sort of possible husbands were stocked just now in London before she selected one.

Elizabeth began her siege of London in the usual manner. She took a furnished house in Mayfair and engaged an excellent chef, for, whatever her particular line should prove to be, those two provisions were essential and must be installed first, just as you put your plumbing into the house before you made it pretty. She had a few friends already living there, one of whom had married a politician, another an artist, another an earl, and she went to dine on the terrace of the House of Commons, and in a Chelsea studio, and in a supremely mournful mansion in Eaton Square. At these houses she met her friends' friends, out of whom she sometimes selected two or three and gave them lifts home in her purring morocco-lined car. She was consequently called on by them, and presently asked them to dinner, and so climbed along in the ordinary way.

But her infallible instinct told her that she was, so to speak, climbing along a horizontal branch of the great tree, and not really making any perpendicular progress. The dinner on the terrace of the House of Commons, for instance, was no good at all; she met some earnest politicians and their even more earnest wives, but there was nothing progressive about it. The dinner in the studio at Chelsea was even worse—it was positively retrograde. Nobody cared about pictures nowadays, nor about the people who painted them, and she very wisely gave nobody a lift home from there, as such people would only impede instead of assisting her. The dinner with the countess in Eaton Square was really little better, for though there were plenty of people there with titles and an air of reserve, they had evidently no more to do with the great social tree than names cut on its bark. Herein she showed a flair vastly superior to many who had set forth on the same brave adventure as she, and who, in their antiquated fashion, thought that rank was in itself an acquisition at your dinner-table. 'For being a duchess,' thought the astute Elizabeth, 'doesn't prevent your being a dowdy.' But the result of these first dinings out was that she was asked to a good many more dinners, and also to subscribe to various charitable objects and take tickets for an immense number of charitable concerts. She disliked

music, but she diligently attended these concerts, and her assiduity
in appearing at them began to give her the reputation of being
passionately devoted to it.

It was as she drove home from one of these melancholy and
melodious gatherings that Elizabeth's great idea struck her.

'I must have a stunt,' she said to herself; 'everyone who gets on has
a stunt. People must get at my house what they don't get
everywhere. What am I to give them?' The point demanded the most
careful consideration. She could give them delicious food, but then
everybody had delicious food nowadays, and there was no credit in
that. She could hire the Russian ballet to come and dance, but then
anybody could see them do that by going to the theatre. She could
get a French actor or two to give small improper plays, but then
nobody understood French, however much they pretended to. She
could give dances, but there were too many of these already. And
then the idea of music, to which she was already supposed to be so
passionately devoted, occurred to her.

Now, such a notion was naturally abhorrent to her, for, as has
been said, she disliked music, and felt restless and uncomfortable
when it was going on, as if a mosquito was buzzing round her head.
But London, she knew, was just now entertaining the remarkable
delusion that it was musical, and the tree-top towards which she was
aspiring was resonant with pianos and songsters. So she determined
to overcome her repugnance to melodious noises or, at any rate,
suffer in silence, and began to think out what manner of music she
would give.

Of course it was no use giving people the music they could get
elsewhere if music was to be the stunt that should hoist her into
Society. The Queen's Hall band performing the overture to the
Meistersingers or the symphonies of Beethoven was not likely to
attract anything except the suburbs, nor was a private ballad concert
likely to further her progress.

And then she suddenly remembered a letter she had received
from a young compatriot of hers who had just arrived in London,
and who was giving a piano recital at the Æolian Hall that evening.
For the purposes of his art he called himself Smirkowski, and in
person he was an extraordinarily handsome young fellow, which
would probably not be a disadvantage. She had not meant to take
any notice of his reminder that he was in London, but now, if his

concert tonight proved to be a success, she determined to engage him to play at her house one evening after dinner. He would be a novelty, anyhow, and London, like ancient Athens, delighted in some new thing, and he had the further advantage of being a foreigner; for London was convinced that no one of English birth could possibly either sing or play a note, whereas all foreigners were musical geniuses.

The critics, however, did not think that this one was, and for once they were unanimous. Elizabeth sent for half a dozen of the daily papers next morning, and all alike agreed that Smirkowski could not play at all. One said that he had not grasped the faintest rudiments of technique, another compared him to a bull in a china shop, another called him a mere impertinence, another recommended him to take to golf, as his only equipment was to hit harder than anybody else. His programme had consisted of classical masterpieces, ballades by Chopin and sonatas by Beethoven, a Bach prelude and fugue, and none of these renderings bore the smallest resemblance to the revered originals. In addition to millions of wrong notes, he took the most unheard-of liberties with his text, and the only thing he could do was to break strings—on this point alone the critics disagreed, for one said he broke three, another that he broke four—while his notion of playing *piano* was to put down the soft pedal and become totally inaudible. His final impertinence was to play the airs of 'Tom Bowling', 'Adeste Fideles', and 'Sally in Our Alley' with original and excruciating harmonies.

And now Elizabeth showed that she was of no common clay. As she read these devasting remarks, her eye brightened, and, instead of feeling discouraged as to the plan she had outlined, she was conscious of a thrill of exhilaration. Smirkowski might not be able to play the piano, but then she did not believe that the sort of folk with whom she desired intimacy knew anything about music. It was clear that he was exuberant and exotic, and as for his playing wrong notes, modern music, which everyone professed to admire so much, consisted of nothing else. She ate a hurried breakfast and went off in her car to the humble address in Bloomsbury from which he had written.

She found him reading his press notices in the deepest dejection. After such universal contempt he had already made up his mind to abandon his second recital altogether.

'Not a single seat will be taken,' he said. 'It's a conspiracy.'

'Well, I'm here to make another conspiracy,' said Elizabeth. 'I want you to play at a party I shall give next week. If that turns out as I expect, I promise to give you plenty more engagements. The only stipulation I make is that you don't play at any other private houses for the next three weeks.'

Smirkowski burst out into a shout of delightful laughter. He was quite unlike the ordinary pianist, being tall, extremely handsome, and short-haired.

'I'll readily promise you that,' he said.

'That's settled, then. As for your next recital, don't abandon it yet. In a fortnight's time things may have happened. And of course you can't continue to live in this terrible place. Come and stay with me.'

At the numerous charitable concerts which she had attended, Elizabeth had met a good many of the musical world of London, and now, without further formality, she asked a dozen of the very best and brightest to dine with her and hear the great new pianist. Some of them had seen the verdict of the press, and thought that there must be something interesting in a man who aroused such virulence of criticism, and those who had not were eager to hear anybody new, and the name Smirkowski was promising. They all accepted, and Elizabeth, with a stroke of genius, had programmes printed with a little nosegay culled from his press notices, for any intelligent amateur would be disposed to think well of a pianist whom professional critics had trampled on. She gave them an admirable dinner, and again, with incomparable wisdom, asked nobody to come in after dinner for the recital. Thus, if Smirkowski was a failure, the disaster and the derision would be limited to a small field, while, if he was a success, all those whom she had not asked would be in a frenzy to hear him, and she could count on bringing together at least one more smart party. Besides, a dozen people reclining in large comfortable seats were much more likely to be appreciative than if they were wedged together in rows of gilt and creaking chairs, and were subject to the interruption of fresh arrivals.

Smirkowski made himself quite charming during dinner. He had no touch of professional pomposity; he was gay, he was handsome, he had exuberantly boyish but excellent manners, and it was with difficulty that he could be disentangled from the clutches of two of the smartest sirens in London and taken to the piano. 'I hope you'll all talk,' he said, as he took his seat, 'and you'll hear less of the horrible noise I am going to make.' Consequently dead silence fell, and Elizabeth whispered a final word in his ear: 'Mind you break some strings,' she said.

Smirkowski opened with a delicately rippling prelude and fugue by Bach, which he turned into a sort of cannibal dance, and Mrs Blankney—one of the sirens—gasped and tied herself into knots. He followed it by a Chopin étude, of which hardly a note was audible, and everyone wore strained concert-faces expressive of esoteric ecstasy. Then came an original setting of 'Rule Britannia', in which he broke two strings. The floor shook with the appalling riot, the windows rattled, and Lady Heycock—the second siren—became slightly hysterical with rapture. When he paused to wipe the legitimate perspiration from his face, she pulled herself together again, and, elbowing Mrs Blankney out of the way, rushed to the piano in order to engage him without delay to play at her Royal dinner-party next week. Elizabeth was close by, and, hearing him— the honourable boy!—express his regrets, could not resist intervening.

'Why, that is a pity!' she said to Lady Heycock. 'Smirkowski is dining here quietly that night, and is going to play his new nocturne to me afterwards—just him and me.'

Lady Heycock gave her the most ingratiating smile. 'Ah, but can't I persuade you to come, too?' she said. 'Just for an hour, dear Mrs Nutter. You will find all the world there.'

Elizabeth knew she was taking a risk, for her mouth watered to accept this invitation, but there was an earnestness in Lady Heycock's voice that boded well. After all, you had to take risks in such a campaign as hers.

'So kind of you,' she said, 'but I never go out after dinner.'

She saw irresolution flicker in Lady Heycock's face. 'But won't you and Mr Smirkowski come and dine?' she asked.

Elizabeth stifled the cry of triumph that rose to her lips, and still hesitated.

'I must ask him,' she said. 'Smirkowski, Lady Heycock wants us to dine with her on Tuesday. Shall we have our quiet little evening another night?'

The news of this wonderful pianist, who bewildered and deafened you, spread like a forest fire, and what added to the excitement about

Smirkowski was that it appeared to be impossible to get him to come and play at your house, for he was always engaged to dine with Elizabeth, and if you wanted to hear him—and not to have heard Smirkowski was to be branded with the sign of Philistia—you had to cadge for an invitation from that delightful Mrs Nutter. This was all magnificent for Elizabeth's purposes, and the cost of listening to his excruciating music every night was really a small price to pay for it. She had, too, to cultivate a face of ecstasy when he was playing, and talk musical jargon. But she studied the language and babbled about syncopated fifths and key-colour, and if she occasionally made a mistake and alluded to Dostoieffski's sixth symphony, there was no harm done. Meantime she and Smirkowski became great friends, and though when they were alone they never talked about music, they thoroughly understood the business side of it. Smirkowski quite saw that he owed the furore about himself to Elizabeth, and she on her side perceived that since she had prevented him, by the terms of their bargain, from playing elsewhere, it was only fair to him that he should have plenty of engagements at her house. That, however, suited them both, for she was raking in all the people she wanted to know. They were both also agreed that he should not play too often, or his boom would collapse. But it had not collapsed when he gave his second recital, for the hall was crammed, and intoxicated admirers presented him with floral offerings, till he looked like a great yellow-haired lamb decked out for sacrifice.

They returned home, when it was over, to a quiet supper, and afterwards, as he wrote his autograph in the albums of numerous young ladies, his jubilant high spirits faded into pensiveness. They were both acutely conscious of the other's presence in a manner they had not previously experienced.

'What's the matter, Teddy?' she asked at length.

He scrawled the final signature and pushed the album away.

'Well, you know, this can't go on,' he said. 'You've often told me that though you adore music, you know nothing whatever about it, and so you must take it from me that I don't know how to play. The critics found that out instantly, and presently everybody else will. I can't play a hang, Elizabeth. I've got impudence and fingers of iron, and that's all.'

'It seems to be enough,' said Elizabeth.

'But they'll find out, I tell you,' he said, 'and it would really be much wiser of me to stop playing before they do. On the other hand, I owe everything to you, and if my playing helps your stunt—why, of course, I'll go on.'

'Teddy, you're a perfect dear,' she said. 'But you've played me into every house in London where I wanted to get, and I owe that to you. And I'm getting on splendidly in my own account now. I always knew I should when once I had a start. I'm just as much a success in my way as you are.'

He gave a sigh of relief. 'That's splendid, then,' he said. 'I won't play to the crowd any more, though, of course, I'll play to you as often as you like. There's another thing too, Elizabeth.'

Elizabeth felt her heart flutter. She knew she was in love with Teddy, and sometimes she thought——

'What's that other thing?' she asked.

He came close to her. 'Why, that I adore you,' he said.

Presently they became sensible again.

'But you mustn't give up your playing,' she said. 'Whether you can play or not, I don't know, but you love it so.'

He looked steadily at her. 'I am going to tell you a secret,' he said. 'I don't love music. I hate music. I should like never to hear another note. It's simply a silly, worrying noise.'

She raised her face to his. 'And didn't you ever guess?' she said. 'I detest music. I can't bear it. Teddy, how perfect!'

AUNTS AND PIANOS

BOBBY DEACON at the age of fifty-five was a very busy young man; no one had so many engagements to little lunch-parties and tea-parties and dinner-parties. He had in fact made a firm rule never to do a crossword puzzle till the evening paper came in, for, if he permitted himself to attack one after lunch, the brisk walk he took every day for the preservation of his youthful figure was sadly curtailed, while if he yielded to temptation after breakfast there was no telling when he would get to his piano, which was the serious work of the day. He played prettily with a butterfly-touch which flitted over the keys with an agreeable lightness, and in his drawing-room, which he called 'the music-room', there were two grand pianos. He had bought one himself, the other had lately been left him by a charming old lady with whom he used to play simple duets. She had in her will expressed a hope that he would often use it, and think of the melodious hours they had spent together. So Bobby, who was of a very affectionate and sentimental nature, constantly played on it, though it was thin and stiff and of a chirping tone, and did not suit the butterfly-touch. In fact, it was very bad for the butterfly-touch, for you had to slap the notes, instead of just alighting on them and sipping their honey. He alluded to the instrument always as 'Aunt Martha', by which endearing title he had always addressed the deceased who, though no relation of his, was thus adopted by affection. Martha had been her Christian name, and 'Aunt' was added out of respect for her superior years. 'I had a good go with Aunt Martha this morning,' was Bobby's bright way of telling his friends at lunch that he had been very industrious.

Bobby lived a good deal among aunts of this description; most of the little lunches and teas and dinners which filled up so large a part of his busy day were taken at the houses of elderly ladies, widows and spinsters for the most part, who called him Bobby or naughty boy. They lived chiefly in the Boltons, or, as Bobby said, the Bromptons, and told him all about their troubles with kitchen-maids, their difficulties in getting the right shades of silk for their embroidery, the aches they had in their venerable joints in this cold weather, and their plans for their holidays. Bobby listened with sympathetic interest to all these deep problems of life, and cheered them up with amusing little bits of gossip from Mayfair, where he often went to tea, and did crossword puzzles with Dowager Countesses. Bobby did not dine or lunch much with those modish dames, but he was in considerable request at tea-time when men were at their clubs or not yet back from their offices. This suited him very well, for though quite sociable, he did not care for the society of his own sex, and felt that they had not much to say to each other, for he took no interest in masculine topics, such as sport and games. Nor did he care for the society of girls, and though he really took great pains in order to interest and amuse them with the little stories the aunts liked so much (though they bridled and called him naughty), sometimes a girl would burst out laughing quite unexpectedly and call him a dear old thing, which he thought singularly irrelevant as well as quite untrue. Or else she would merely yawn, which was rude. . . . Girls all seemed to him to be indistinguishable from each other with their cropped hair, and long sunburned-stockinged legs. . . .

'And it's even worse if they've got hats on,' said Bobby to Mrs Fakenham, who had lately become Aunt Judy, 'and then I never recognize them at all. If men wore on their heads scarlet straw waste-paper baskets which came down to the end of their noses, and painted their lips the colour of their hats, and never took a cigarette out of their mouths, they wouldn't expect to be recognized. But girls do expect me to recognize them, Aunt Judy. I can't bother myself with them.'

Aunt Judy sneezed several times, for she had a frantic cold in her head, and then held up a reproving forefinger.

'You're a very spoilt boy,' she said wheezily. 'You don't take any trouble with girls.'

'Indeed I do,' said Bobby, 'but I never know who they are, or what to talk to them about. Even if they are pretty I can't see their faces, and when they take their hats off, I think they're boys, and boys are odious. I'm glad there are no boys or girls here tonight.'

Aunt Judy had distinct remains of coquettishness about her.

'Nonsense, Bobby,' she said. 'Fancy your pretending that you prefer to come and dine alone with an old woman like me to having some pretty girl to take in to dinner. But I thought I would be selfish tonight, and have you all to myself. Next Thursday, when you're dining with me again, there will be a little party, and you'll have to be host at dinner, and play to us afterwards.'

'I shall enjoy tonight most,' said Bobby, who always flattered his aunts.

Aunt Judy, as usual, gave him a remarkably good dinner, and Bobby, who was greedy, liked that. A crossword puzzle succeeded, and as he had done a little work at it already (though it was not necessary to mention that), he was very brilliant about it. Then Aunt Judy thought she would like to try Brahms's Hungarian dances arranged for four hands, and of course Bobby had to say that he positively preferred playing the bass, which was tactful though false. Indeed, playing the piano at all in Aunt Judy's house was always a trial, for the instrument was an antique Hobbner concert-grand of ear-splitting quality with two notes in the bass that stuck when struck and wouldn't leave off sounding, and one that produced no sound at all. And playing with Aunt Judy did not mend matters, for she filled up most of the seating capacity opposite the keyboard, and Bobby was squeezed away at the lower end of it, where he could hardly see the music. When in action, Aunt Judy kept her foot firmly on the loud pedal, counted in a hoarse voice, and corrected her own wrong notes, so that her counting got out, and it was impossible to tell where she was. But she enjoyed it immensely and at the end of a strenuous hour, patted his hand, and said it was a pleasure to play with him.

Bobby edged himself away when it was over, and Aunt Judy, as usual, with all the piano at her disposal, played the short prelude by Chopin with the slow big chords.

'Splendid!' said Bobby. 'You played that divinely.'

She played it again and rose.

'I am proud of my piano,' she said. 'Magnificent tone, is it not? My dear husband chose it for me, and it was always reckoned a very fine instrument. Now we'll have a little gossip over the fire while you drink your whisky and soda. Dear me, what a dreadful cold I've got! And there are a hundred things I must do tomorrow.'

'If I were you I should stop at home and nurse it in this bitter weather,' said Bobby.

'Not at all!' said Aunt Judy. 'Disregard a cold: that's the way to get rid of it.'

Bobby also was full of engagements next day. He went to lunch with Mrs Trask (Aunt Fanny), who in her faint and remote voice sang 'Ich Grolle Nicht', and several other courageous songs, which he accompanied for her on her new and rather unripe Boddington grand. She was the most devoted of all his aunts, and the richest and the least robust, and sometimes Bobby when he was not thinking what he was thinking about allowed himself to think about Aunt Fanny's entire lack of relations. ... When Aunt Fanny had finished singing, she sank, rather exhausted, on to a small chair encrusted with mother-of-pearl and tortoise-shell. Aunt Fanny was very small and all her furniture with the exception of her grand piano was small and fragile, and Bobby trembled to think how awful would have been the crash if Aunt Judy had sunk exhausted on that chair, or indeed anywhere in Aunt Fanny's house except on the floor. But that was not likely to happen, for he never brought the aunts into contact with each other, feeling, vaguely but correctly, that each individual aunt would take less tender interest in him, if she knew there were others.

'No one sings "Ich Grolle Nicht" so beautifully as you, Aunt Fanny,' he said. 'I can't think why it should be considered only a man's song. You make it so deliciously feminine without losing any of its bravery. And your new piano is quite splendid. You were lucky to get such a beauty. It will improve too.'

'I'm glad you like it,' said Aunt Fanny. 'It wants playing on, but I think it will be satisfactory.'

Bobby applied the butterfly-touch to the keys, and got a fair effect.

'Satisfactory! I should think so,' he said. 'And as for its wanting to be played on, there's somebody who wants to play on it whenever he's allowed.'

Bobby slid into a fluid little morsel by Debussy, which was a favourite of Aunt Fanny's, though Aunt Judy declared that it sounded to her like a child whimpering next door. It made Aunt

Fanny feel ill and unhappy, which she liked, for one of the greatest
joys of her life was feeling ill and thinking she was going to die. She
gave a wan little smile when Bobby had finished.

'So sweet and miserable!' she said. 'Thank you, dear. And now,
Bobby, I've got something to tell you. Perhaps you wondered why I
sang those songs just now. It was to make myself brave. I've got to be
brave.'

'Dear Aunt Fanny, what's the matter?' asked Bobby. 'You're not
ill, not particularly ill, I mean?'

'Not worse than usual,' she said. 'But my doctor has told me that it
would be most unwise of me—suicidal, indeed, he said, when I
pressed him—to spend the winter in England. I've got to go south,
and I shall let this house for six months. A terrible wrench. And
you'll miss me, dear, I know.'

'It will be horrid without you,' said Bobby warmly.

Aunt Fanny sighed.

'I dare say I shall never see England again,' she said, 'and I think it
is only wise to make all arrangements before I go. I shall let this
house, as I said; indeed, I've got a very good offer for it, but I don't
want to leave my new piano to be strummed on. So will you house it
for me, Bobby, while I'm away? I remember your saying that your
music-room was often so cold, and I thought you would like it in
your dear little sitting-room by the front door.'

It seemed to Bobby that Aunt Fanny might store it, for he didn't
want it at all, but it would never do to reject any kind idea of Aunt
Fanny's. It was much wiser to accept it with alacrity and enthusiasm.

'Oh, that is good of you, Aunt Fanny,' he said. 'Fancy my having
that delicious piano to play on cosily in my little den. Every time I
play it I shall think of you.'

'I thought you would like it,' she said. 'And I shall think of you
playing on it when I am far away. And if I never come back, as seems
likely, I have left it you in my will.'

She gave a little gulp.

'You will have many happy years to play on it after I am gone,
Bobby,' she said.

She talked about her journey a little more, as if it had been the
conveyance of a corpse to its final resting-place. Bobby cheered her
up, and, since Aunt Fanny made no suggestion of sending the piano
to him, it was settled that he should notify the makers that it was to

be taken to his house next week on the day that Aunt Fanny started for the Riviera. After that it was time for Bobby to hurry off to Curzon Street for tea, where he did a crossword puzzle with a Marchioness.

Bobby dined that night with Miss Pattison, who was not old enough to be his aunt, for she was only a year or two his senior, and therefore was Cousin Ella. It was always lively at Cousin Ella's, though she occasionally frightened Bobby by telling him that he ought to marry, and then casting her eyes modestly down. But she had said so with such frequency that Bobby had begun to think that she didn't really mean *that*; besides, if she had intended to marry him she would probably have proposed to him by now. He preferred, however, that there should be other guests at Cousin Ella's, but tonight, as last night with Aunt Judy, he was alone with her. As usual there was music, for Cousin Ella had a set of jazz-band instruments which she played fortissimo while Bobby thumped her Boddington boudoir-grand and tried in vain to make himself heard. When they were both thoroughly exhausted they sat over the fire, Cousin Ella in a very low chair, showing an incredible length of brawny leg.

Cousin Ella often gave Bobby delightful presents, for like most of the ladies he had chosen as aunts and cousins, she was very well off. Last Christmas she had given him a beautiful fur coat and on his last birthday, now nearly a year ago, a handsome Chippendale mirror, and books and flowers were constantly showered on him; indeed, he seldom left the house without something pretty or useful. She lived in Cromwell Road, but, except for a party, never used the big rooms upstairs, preferring the cosiness of the little sitting-room on the ground floor, where they sat now.

'I'm going to make a change in this room, Bobby,' she said. 'I'm going to hoof out the boudoir-grand and have a cottage piano. Makes more room.'

'Oh, don't do that,' said Bobby, 'it's such a delicious one. Or perhaps you're meaning to have it upstairs.'

'No, I shan't do that,' said Cousin Ella, 'I don't want a piano there. I'll tell you what I'm going to do with it. It's somebody's birthday next week, Bobby!'

Bobby had a feeling, like that in dreams, when the sleeper is aware that a nightmare is coming.

'Yes?' he faltered.

'Well, it's my birthday present to you, Bobby. You always tell me what a lovely piano it is, and I shall love giving it you. There! Send round for it as soon as you like, for my new one is coming next week. Such a joy it will be to think of your playing on it.'

'Cousin Ella, it's too good of you,' began Bobby. 'But——'

'There isn't a "but",' said Cousin Ella. 'I'm determined you shall have it. Don't thank me, dear: I love giving it you, for I know how you'll appreciate it. Order a van from Boddington's and it will be ready for you.'

Bobby wrote to Boddington's next morning, asking that a van should be sent on Wednesday next, first to Aunt Fanny's to fetch her piano to his house, and then to Cousin Ella's, to fetch hers. He had just written this, when he was rung up from Aunt Judy's to be told that her dinner was put off as she was seriously ill with double pneumonia. . . . Poor Aunt Judy died on Saturday, and the funeral was fixed for the following Wednesday. Bobby was very much distressed and ordered some black clothes in a great hurry.

He went to the funeral on Wednesday, leaving word with his parlour-maid that two grand pianos would presently arrive, of which the larger was to be put in the little sitting-room by the front door, and the smaller in the music-room. He thanked Heaven that the latter was only a boudoir-grand, but, even as it was, so much furniture had to be moved out of the two rooms into the dining-room that it was difficult to see how food could possibly be dispensed there. But he could sell some of his larger pieces, and though four pianos was in excess of bare musical requirements, Bobby's sentimental nature still went out in gratitude to the kind donors of these instruments. He might, of course, also sell his own admirable Bilhausen, but he did not want to do that, since it was far the best of the lot, so it and Aunt Martha and Cousin Ella would be fitted into the music-room, and Aunt Fanny into the little sitting-room. And he hurried away to Golder's Green.

A light drizzle was falling, and when he came back, cold and damp and depressed from the last sad rites, it was clear that troublesome things were happening. On the pavement outside his door, shielded by a tarpaulin from the inclemencies of the weather, stood the smaller Cousin Ella, while the larger Aunt Fanny completely blocked

the narrow passage within. By no sleight of hand, so the pessimistic
foreman told him, could she possibly be introduced into the little
room for which she was intended unless the window were taken out,
the area-railings pulled up, and a crane erected to swing her in. Even
then this engineering feat seemed highly risky, for the balcony of the
room above, where the crane must be placed, would probably give
way, and Aunt Fanny be precipitated into the area, with or without
loss of life, and completely block up the entrance to the coal-cellar.
It would be possible, however, though difficult, to entice Aunt Fanny
into the music-room, and get Cousin Ella into the little sitting-room.
About nightfall this was accomplished, and Bobby paid an
outrageous cheque to the foreman.

Dining-room, music-room, and sitting-room were now tightly
packed. It was perhaps possible for a slim man like Bobby to get
access to the keyboard of any of the three gigantic instruments which
filled the music-room, but no fire could possibly be lit there, since
Aunt Fanny's thin end projected over the grate, and the idea of
holding any little musical party there again was simply laughable.
Bobby's small sitting-room was also quite ruined, for Cousin Ella had
crowded out the sofa and the writing-table, while the dining-room
was like a well-stocked furniture shop. Bobby ate a miserable dinner
under the disapproving eye of his parlour-maid, who looked as if she
was going to give notice, and wandered from room to room
disconsolately, unable to think of any plan which would render any of
them habitable.

The smart rap of the delivery of the nine-o'clock post gave him a
slight ray of comfort: there would probably be some pleasant
invitations to lunch or dinner. . . . He went to the door, but found in
his letter-box only a long envelope with a typewritten address, which
looked as if it might be connected with taxes. He opened it, and
found it was from Aunt Judy's lawyer.

'The late Mrs Fakenham,' he read, 'has left you in her will her
grand piano by Hobbner, with a touching and affectionate message,
expressing the hope that you will spend many pleasant hours in
playing it with kind thoughts of the donor. As there is to be a sale in
the house almost immediately, we should esteem it a favour if you
would make arrangements for transporting the instrument to your
own residence as soon as possible. . . .'

THE GUARDIAN ANGEL

MRS ATTWOOD was an exceedingly good-natured woman, and her time and her advice were always at the service of her friends in the cause of their peace, happiness, and prosperity, and no doubt her money would have been at their disposal also if she had not wanted it all for herself. It was not because she lived extravagantly or self-indulgently that there was none to spare, but because there was very little of it, and it was only by the most careful economy that she managed to keep herself so nicely dressed and live in her pretty little house in Kensington Crescent. It was a long way off from the more exalted squares and streets of Mayfair where she so constantly dined and lunched, but people were kind, and often sent a motor for her, or, if this did not occur to them, she ascended an omnibus or descended into a tube with the utmost agility and cheerfulness. Even so she could not have managed to live in London at all if she had not always let her house for a couple of months every year, with everything convenient and complete, from servants who welcomed her tenants with smiling faces, down to notepaper with the address and telephone number printed on it, in the stationery-cases. Little things like that always gave satisfaction, and anyone in want of a small residence ready to step into could not but be charmed by the dainty freshness and admirable appointments of her rooms. The telephone had an upstairs extension, the bathroom had a hot towel-rail, the kitchen was white-tiled and spotless, and an unlimited supply of hot water was furnished by a furnace that prospered under the most meagre and inexpensive diet. It had meant a considerable outlay to fit up her house so perfectly, but the money had been excellently invested, for while rows of mansions stood

gauntly hungering for tenants, Mrs Attwood was always sure of a let
for any period: she had even once let her house in August. During
these tenancies she retired to lodgings in Brighton and came back
much refreshed by the bracing air and ready for anything.

She was returning home one afternoon in early April from a small
luncheon-party, and was full of a glow of general appreciation. There
had been a fortnight of foggy and dispiriting weather, and the gay
brilliance of sun and promise of spring was much to her mind. She
had appreciated her lunch (for she was not so narrow-minded as not
to include exquisite food among the blessings of life), she had
appreciated eating at Lady Rye's house with a small company of very
distinguished people, and, above all, she had appreciated the fact
that it was she who was the real originator of this remarkable little
feast. For two friends of hers, Daisy Bright and Marjorie Yorkshire,
had had, to put it mildly, the devil's own row over that foolish game
of bridge, and had accused each other of being swindlers and
sharpers. Having utterly blackened each other's characters, yet still
lusting for wider vengeance, they had proceeded volubly to blacken
the characters of all each other's friends, and as Mrs Attwood was a
friend of both, some very surprising things were said about her.
Something had to be done, and instead of making matters worse for
everybody, including herself, by any sort of retaliation, she had gone
with her winning manner to Lady Rye.

'Darling Peggy,' she said, 'I'm going to ask you to do a very kind
and dear thing. (Because you are such a dear kind thing.) Such a row
going on between Daisy Bright and Marjorie Yorkshire. You've heard
about it?'

'I've heard about nothing else for ages,' said Lady Rye.

'Well, I want you to stop it. A woman like you, with your position
(I always say that nobody has ever been on such a pinnacle as you),
can do it, and only two or three women in London could.'

Lady Rye laughed.

'I should be delighted to stop it,' she said, 'because I'm told that
Marjorie has been saying that I habitually drink too much. But how?'

'Ask them both to lunch, without telling either that the other is
coming, and get a few of the very brightest and best. Tell them there
will be bridge afterwards, and then put them at the same table.
They'll cave in; they won't dare to do otherwise in your house.'

'But, suppose they don't cave in? Suppose they throw the cards at

each other? They're wild cats, they're Habbakuks; they're capable of
almost anything.'

'Darling, trust me. I'm a major prophet. I know I'm right. I know
what you can do.'

Lady Rye liked being told she was on a pinnacle. Any woman
would.

'I'm terrified, but I'll try,' she said. 'I only stipulate that you come,
and if anything goes wrong I shall say it's your fault.'

Of course nothing had gone wrong: the wild cats had caved in, and
purred, and been partners, and rooked everybody. What would have
happened if Providence had not given them marvellous cards and
they had lost, need not be enquired into. But Providence had done
the right thing, and Constance Attwood had done the right thing,
and felt as if she was the Guardian Angel of Society. In fact, as she
walked back to Kensington Crescent with well-earned complacency,
she thought Guardian Angel would be quite a good name for her.
She would tell her friends that she had heard that people were
calling her 'Guardian Angel', and they would spread it, and call her
G.A. for short.

Among the guests at this Peace Conference had been Mr
Ferdinand Close, millionaire by birth, and by calling the most hotly
pursued of middle-aged bachelors, and G.A. had long turned a
pensive eye on him, cogitating plans for his welfare and happiness. It
must instantly be premised that she had now no notion of personally
partaking in his welfare and happiness, for she had done her best to
involve herself in it and to put an end to her economies and her
widowhood; but, in the odious slang of the day, which she never
used, she had convinced herself that there was 'nothing doing'. But,
such was her great good nature, she bore him no grudge for his
obduracy, and was just as desirous to secure his happiness as ever. Of
course she would score by it, for if through her agency and abettings
some friend of hers became Mrs Close, she would naturally be
grateful, and though G.A. was most modest in her requirements, she
knew that no woman could have too many friends with beautiful
places in the country and moors in Scotland, and houses on the
Riviera. She hated Scotland, but she liked grouse and the Riviera.
Indeed, as she often told herself, she did not even want to be very
rich herself, provided she had an ample bodyguard of very wealthy
friends who were devoted to her. That was a more altruistic attitude,

and it saved trouble: you got the winter sunshine and the grouse, without expense. She let herself into her house, and went straight to the table where lay the tablets on which were written the messages of those who had yearned for her on the telephone since she went out. That ranked high among the minor pleasurable anticipations of the day, especially when (as this afternoon) she had at present no engagement for the evening, and might have to spend it alone without more diversion than 'something on a tray' and that very clever book of Mr O. Sitwell's would afford her. The first message began, 'If Mrs Attwood is free tonight, would she——' Mrs Attwood settled that she would, whoever it was and whatever it was, for she ranked human intercourse higher than any literary treats, and put on her glasses to read the rest, since the fate of being alone that night was now averted.

'Dine,' so the message went on, 'quite alone with Mrs Soningsby. She very much hopes Mrs Attwood is free, as she wants to see her particularly.'

G.A. instantly tinkled for Julia Soningsby. She had a notion of what she might perhaps want to see her about, and though it would be very sad if she was right, it would also be very interesting, and eventually something suitable might come of it. But Julia was out, and she could learn no more for the present.

There were other requests for her as well, and, having consulted her engagement-book, she left her maid to accept most of them, and went upstairs, thinking violently about Julia. Of course, she was exactly the wife for Ferdinand Close; she was extraordinarily pretty, still quite young, and possessed of wonderful charm. The only objection was that she was already married, but in this S.O.S. call Mrs Attwood saw a gleam of hope. She and Geoffrey Soningsby had made a very poor job of their marriage: they were always having quarrels and disturbances, and they were both so fascinating that it had long seemed a dreadful thing that so much enjoyable material should be wickedly wasted like this. They had made a mistake in marrying each other, and G.A. almost hoped that they had come to the only sensible conclusion and had settled to free themselves. Two years were surely long enough to spend in finding out that they were incompatible, and it was high time that they gave each other a chance of happiness. 'It is no true marriage,' she thought, 'when there is so much misery.'

She turned to her tea and her letters. The very first letter she opened was a respectful communication from a house-agent asking if she would consider letting her house for two months from the middle of April till the middle of June. This most desirable client who wanted it had already been a tenant of hers, and would be pleased to give the same rent as before. Mrs Attwood was sorry this application had come just now, for she would have preferred to spend the next two months in London herself, and her first impulse was to refuse, and chance getting a let for two of the autumn months. On the other hand, she wanted to build out a scullery with a bathroom on the top of it, and an eight weeks' rent at twenty-five guineas would enable her to do this without encroaching on her capital. She was in two minds about it, which was a rare thing for her, and by way of helping herself to come to a decision, wrote to the agent to say that she would let her house for the period required at forty guineas a week. That would be too good a let to refuse if the desirable client rose to it; besides, she had long thought that, considering the paper in the stationery-cases, and the white tiles in the kitchen, her house ought to command a higher price. There was the inconvenience to herself also: usually she let when she did not care about being in London, but now she would much sooner remain here. 'So the desirable client must pay for it, if he wants it,' said G.A. thoughtfully as she went to dress.

Julia came downstairs some ten minutes after G.A. had arrived at her house, looking like a handsome boy who had chosen to cut his hair shorter than usual, and to wear a sapphire-coloured sack which very nearly came to his knees. She began to talk slowly and softly before she had really entered the room. Apparently, they were to dine there, for there was a small table laid with knives and forks, and ash-trays.

'It's all been so hopeless for a long time,' said Julia. 'I adore Geoffrey, and he adores me, but what's the use when we quarrel all the time? So heartrending when one's always furious with the man one adores. Much better to give it up, and marry other people. Then we can meet and be devoted to each other without getting on each other's nerves. I'm miserable, and so is he; but it's the only way out.'

Julia always made Mrs Attwood feel Victorian: she felt even more so as the meal, which consisted chiefly of caviare and cocktails, proceeded.

'We went down together to Roehampton this afternoon,' said Julia. 'Geoff was playing for the home team, and we had the most awful row. I began, I'm bound to say, for I asked him who was paying for his ponies, and he, with a horrid sneer, really horrid, asked who was paying for my frocks. Of course, nobody is paying for either: so silly. Just bills. And then I saw that he meant to insult me by asking who was paying for my frocks. Naturally, I told him that he must instantly apologize for that, and he said he'd be damned first, and that my asking who paid for his ponies was just as bad. We got nastier and nastier, and at last we settled we couldn't stand each other any longer. This row was the last straw, and it put the lid on, if you see what I mean. And it was such a relief: we became friendly again at once. It was the first thing we had agreed about for weeks. But we thought we had better ask somebody's advice first, and we both chose you. So I rang you up as soon as we got to Roehampton. Geoffrey's going to stay at his club for the present, and I'm to stop here. We tossed for that: I was quite ready to stay at my club. We're not to be divorced at once, but the moment either of us wants to marry somebody else, he's going to write to me to say he's never coming back, and tell me the hotel he has been at with Her. I think it's sweet of him to take all that part of it, because he'll hate it. So there it is, and if I was your daughter, darling Connie, wouldn't you advise me to do that?'

'Too tragic,' she said; 'but I do think you're right.'

'I'm glad of that; but it isn't tragic. It's only sensible to repair a great mistake. Whom shall I marry next? Do find someone for me as quick as you can, for the sooner this stage is over the better. I want somebody rich, dear Connie, because half the rows between Geoffrey and me began over miserable little sums of money. We borrowed taxi-fares from each other and tried not to repay them, and so that led on to violent criticisms about character. So sordid: nothing spoils romance so utterly as sordidness. And I don't think I want to try romance again. Safer without, isn't it? Besides, I gave Geoff all the romance I had got. Some middle-aged man, don't you think? There must be lots of them. . . . Oh, wait a moment; do you mind while I ring Geoff up. I said I would let him know at once what you advised.'

Since Julia was anxious to find somebody 'to marry next', as soon as she possibly could, it was easy to introduce the name of Mr Close. Julia vaguely remembered him: he was 'square and kind'? And if it was certain that he was rich, she would give him her careful consideration.

'Ask him to tea or lunch, darling Connie,' said Julia, 'and let me come. How about tomorrow? You're a perfect fairy-godmother, darling Connie: you're a guardian angel.'

'That's what they all call me, I hear,' said Mrs Attwood, 'and I'm sure an old woman like me couldn't have a nicer nickname.'

'It isn't a nickname,' said Julia. 'It's you. And you shall be G.A. for short.'

The desirable client of the house-agent's tried a little futile bargaining, but G.A. felt sure he meant business, and was firm, with the result that within a day or two he capitulated.

Mindful of her mission as guardian angel, she made it her first care to contrive meetings between Mr Close and the bride she had ordained for him. There were several such: first the original tea-party at her house, then a bridge-party at each of the wild cats', and a dance at Peggy Rye's. This was on G.A.'s last night in London, and she passed them sitting out together on the stairs, and neither took the slightest notice of her. She left town next morning, confident of success. Perhaps she would not have been quite so pleased with the prospect if she had known that Geoffrey and Julia dined together a few nights afterwards with Mr Close, and that Julia repeatedly called him Uncle Ferdy.

Luckily for her sense of success, G.A. knew nothing about that, and settled down into her banishment at Brighton with the cheering consciousness that all went well with her guardianship, and that forty guineas a week was blessedly oozing into her bank balance from Kensington Crescent. But this gathering affluence had no effect on her modest mode of life, for it was all earmarked for scullery and new bathroom, and she continued to occupy back bedroom and front sitting-room in her familiar lodging. She did not pay inclusive *pension* rate, but a most reasonable charge for room and breakfast, with the undertaking that when she wanted to lunch or dine at home, she should order her provender and pay a small percentage for the cooking of it. On the whole, this was the most economical plan, for though she was banished from London, fragments of London came down to Brighton for long weekends, and from Friday till Monday there were often friends of hers at the Hampden Hotel who invited her to dine. She was well known there now, and usually spent the hour of tea-time there in the gorgeous lounge, looking at the papers and the afternoon arrivals. She was privileged to do that without ordering anything: the head waiter was aware that she constantly sprang up with a glad cry as the weekend visitors came in, and kissed the wearer of the most sumptuous sable coat. The clerk in the bureau knew her also, and the visitors' book in which they recorded their names on arrival was always open to her inspection. Generally there was someone on whom she left a card with her address, or to whom she wrote a scribbled note.

Her visits here, productive of many pleasant little lunches and dinners, were interrupted after ten days because she caught a really shocking cold and stayed indoors. She thus missed a whole weekend of visitors, and it was not till Tuesday in the following week that she took up her accustomed place under the pretty palm-tree in the centre of the lounge, which commanded a good view of the revolving door of entrance. But there was nothing here to-day that looked in the least like an invitation to dinner, and presently she rose and went out. The visitors' book was alluringly open in the bar-window of the bureau, and it would be amusing to see the names of all these odd folk. She turned back a couple of pages to run through the entries of those who had been here for the weekend, and then emitted a shrill unstifled exclamation as she read, 'Mr and Mrs Geoffrey Soningsby, London.' They had come last Friday, and they had stayed till this morning.

Her first impulse was one of pious thankfulness that she had not encountered poor Geoffrey and been obliged to cut him; her second, of impious curiosity as to what his companion was like. She stifled the second instantly, as being prurient and unbecoming to a guardian angel, and hurrying back to Montpellier Terrace, feverishly considered what she should do next. No doubt Geoffrey would let Julia know through the medium of some hard, formal, legally phrased letter what he had done, and it would be dreadful for her to hear about it like that. She was still very fond of him, though she had decided so rightly to live without him, and it was clearly in the province of a guardian angel to let the news reach her in less brutal a manner. . . . And then there was another consideration: though this fatal and final act on the part of Geoffrey seemed at first to imply that Julia had asked him to give her release (for they had decided not to be divorced till one of them wanted to marry somebody else), it did not prove that Julia had made up her mind to marry Mr Close. It might have been Geoffrey who sought his release because he wanted to marry again. If that was so, it was most crude of him to be so quick about it, and poor Julia would feel it terribly. That she was possibly being crude, too, if she contemplated immediate matrimony, did not occur to her guardian angel, and she was merely wildly excited at the thought that she and Mr Close had arranged it so speedily. But why had not Julia written to her angel to tell her? Not one word had the angel heard since the Brighton banishment began. It must be Geoffrey who wanted to marry again, and though she thoroughly despised curiosity, Mrs Attwood felt that she must know all about it at once. Besides, Julia might be needing her comforting presence, if she already knew. G.A. sincerely hoped she had been spared that shock, and determined to go up to London at once and administer it firmly but lovingly.

She had telegraphed to Julia that she was coming to see her on an urgent errand, and went straight to her house. She was in her bath, but would be only a few minutes, and while she waited G.A.'s observant eye thrilled to see that there were signs of a man's presence in the little sitting-room: a pipe lay on the chimney-piece, a tobacco pouch had evidently fallen from it on to the hearth-rug. . . . And then Julia entered radiant and dressing-gowned, and fragrant with bath-salts. . . .

'And what is it?' said Julia.

'Darling, I know you're fond of Geoffrey still,' she said, 'and what I've got to tell you may hurt you. So I came up from Brighton to tell you myself. He has done it: he has been at the Hampden Hotel with Her. I saw the names in the visitors' book, "Mr and Mrs Geoffrey Soningsby." From Friday last till this morning.'

'Quite a little honeymoon,' said Julia.

For the moment that struck G.A. as terribly bitter, but then Julia began to laugh, and the more she laughed the more completely puzzled grew G.A. When she had enjoyed this incomprehensible joke to the full, she gave explanations, punctuated by hoots and giggles.

'Oh, my dear, how funny,' she said. 'It was all Uncle Ferdy's doing.
. . . Of course, you've been away and have heard nothing about it . . .
He said it was nonsense that two people like Geoffrey and me, who
really adored each other, couldn't hit it off. It only wanted a little
tact and a little money. . . . So he gave Geoffrey a place in his
business, and talked to us (such an old dear) about face. Oh, and he
was far from pleased with you. . . . I'm so sorry about that. . . . He
said you ought to have taken exactly the opposite line, and I'm afraid
I think he was right. . . . But you meant so well, didn't you, G.A.?'

'Then Mr and Mrs Geoffrey Soningsby——' began G.A.

'Yes, of course, Geoffrey and I. We had a lovely time. A little
honeymoon over again. Ever so sweet.'

THE QUEEN OF THE SPA

MISS JESSICA WINTHROP was taking her tea, on this warm afternoon, at one of the little iron tables which fringed the lawn in front of the hotel. This was a post with strategic advantages: she could drop bright little words of encouragement to anyone who was playing on the putting-course which zigzagged over the lawn, and she could also make an early inspection of any arrivals by the train from London which was due about this time. If the weather was inclement she took her tea in the hall and observed them from there as they passed through to see their rooms. As soon as they had gone she looked at the names they had signed in the visitors' book.

Miss Winthrop was a very regular visitor to this Crown Hotel at Newton Spa; she much preferred it to the larger establishment, the Royal, and the Howard Arms which seemed to her garish and noisy and expensive. The Crown was a more friendly and home-like place; they were, as she said, quite a happy family at the Crown, and she was undoubtedly the head of the family. She made friends at once with newcomers (or, if they did not appreciate her attentions, enemies); she had little conversations with the head-waiter, so that she was never given the leg of a chicken, but always the wing; she got up bridge-tournaments and putting tournaments and games in the evening. They were not romping-games (because most of the visitors suffered from rheumatic troubles, which they came to Newton Spa to get rid of, and hobbled or limped or had stiff backs or shoulders) but sedentary, amusing games like Consequences and thinking of a word while somebody limped away out of earshot from the lounge, and then was recalled and made to guess it. If she decreed an evening of bridge, she tripped round to the other tables when she was dummy to see how other people were getting on, and she had her particular armchair in the cosiest corner of the lounge when the weather was cold, and by the window when the weather was sultry, which was recognized as peculiarly hers. Occasionally a newcomer, ignorant of its sacred character, ventured to occupy it, but on Miss Winthrop's appearance the hall-porter whispered a word or two to the intruder.... But sooner or later in the evening, whatever diversion she had decreed, she was induced to sit down at the piano and play some sweet slow movement by Beethoven, or one of the less agile preludes of Chopin or the 'Largo' by Handel. Indeed, the music-stool was as much her throne as the particular armchair, and no one ever touched the piano when Miss Winthrop was present, except by her special request. When she, however, was induced to favour the company, all conversation ceased or was conducted only in the discreetest of whispers. If it continued, Miss Winthrop ceased until it did. But this rarely happened: an automatic hush fell on the lounge at the first firm touch of her rather knobby hands. ...

She had only just poured out her first cup of tea when Mr Foster came out on to the lawn with his putter. Mr Foster, a stout, middle-aged clergyman, was a great favourite of hers, and acted as her lieutenant in getting up diversions for the happy family. He was expecting a friend to come out for a match presently, and till then he practised, while Jessica watched him with advice and applause.

'Oh, padre, that was a good putt,' she said. 'I shall never dare to play you again if you putt like that. ... Ah! Now you've done what you told me never to do: you were two yards short that time. Never up, never in: how that sums it up! Oh, but you've holed it, so you get your two. And here's Mr Leader: now I shall enjoy seeing a match between you two.'

Mr Leader was a very different person from the padre. He was a gruff, unsociable sort of man, who had been known to refuse to go out of the lounge while they thought of a word, and once when he did go, he went to the smoking-room (for Jessica discouraged smoking in the hall) and refused to come back. But he was a newcomer, and she still hoped that he would fall into line. Just as they began their athletic tussle, the hotel bus came back from the station, laden with an immense quantity of luggage, and there got out a tall stout woman, followed by another who was evidently her maid and carried a cushion and a rug and a jewel-case. Somehow Jessica distrusted the arrival: she was terribly smartly dressed with skirts that, considering her build, were unnecessarily short. She was smoking also, and that argued ill, and her immense quantity of luggage argued ill, and her maid argued ill. She might be nice enough, thought Jessica, but so much pomp was not quite the right note at the Crown. . . . Presently she would go to the visitors' book and see who this was.

She turned her rather distracted attention to the putting-match. The padre was in wonderful form, and Mr Leader was getting grumpier and grumpier, in spite of Jessica's encouragement, which she showered on him for propitiatory reasons.

'Oh, Mr Leader, that was hard luck!' she said. 'The naughty ball! It ought to have gone in. And then the padre goes and lays you a stymie. No one can play against such bad luck. . . . There, that was a beautiful putt of yours. Well I never! If the padre hasn't holed out in one. . . .'

Suddenly her stream of encouragement ceased, for from the open window of the lounge there came out the sound of brilliant roulades from the piano. It was some dreadful piece of ragtime music—which Jessica detested—all execution, and twirls and shakes and octaves. She sprang up, and nearly trod on Mr Leader's ball.

'Who can that be?' she said.

'Hi! Fore!' said Mr Leader. 'Just at your feet. . . . I shouldn't wonder if it was my sister-in-law. I believe she was to arrive today. Now we shall get some music.'

Jessica hurried indoors with this blasphemy in her ears, and only pausing to see in the visitors' book that it was Mrs Leader who had arrived just now, went into the lounge.

There, without doubt, at the piano was the woman who had filled her with instinctive distrust, fireworking away all over the keys, with a cigarette in her mouth. . . . With only one moment for consideration, Jessica made up her mind what line to take, and sat down very stealthily nearest the piano, and assumed an expression of delighted gaiety. She gave little smiles and nods of her head in time to the music, and when the last distasteful chords had been played, she turned with her most winning expression to the pianist.

'Delicious!' she said. 'Oh, what fun! Makes me want to dance. *Thank* you.'

Mrs Leader was unaware that she was being complimented by the Queen of the Crown, and saw in her only a rather ridiculous old thing with forced smiles on her acid face. As for her wanting to dance . . .

'Nice of you,' she said. 'What a foul piano!'

Now Jessica had chosen that piano at the request of the manager some seven years ago. Her smile became a little wintry, like a gleam of sun on a day of north-easterly gale, and she reconsidered her policy.

'I am sorry you find it so,' she said. 'Perhaps I am accustomed to it, for I like the touch. *Please* go on playing to me. I adore music.'

Mrs Leader had already risen.

'No, it's your turn,' she said quite amicably.

Jessica slid on to the music-stool. After all, it was her throne.

'Terribly out of practice,' she said. 'A little Beethoven? Or Chopin? Or Handel? The "Largo"?'

'Don't think I know it,' said Mrs Leader.

Jessica made a pained face, which she dexterously transformed into one of pleasurable anticipation, and after that into a rapt expression, as she struck the first chord, of musical absorption. She closed her eyes, as her habit was.

Mrs Leader found that she did know it, but that she didn't like it. Besides, the woman couldn't play at all, and she wanted her tea. She did not mean to be rude, but as this famous piece went at a funereal pace, and the pianist's eyes were closed, she thought there would be time to slip into the hall, order her tea, and come back before it was over. But from the hall-door she saw her brother-in-law putting on the lawn, and forgetting about the musical treat inside, went out to greet him. Thus when Jessica, after dwelling on the last chord, opened her eyes again, she found herself alone, and felt sure that her instinctive distrust had been only too well founded. These dark forebodings—she was never wrong in her first impressions—were speedily and amply confirmed.

The visitors at the Crown were of a punctual habit with regard to dinner, and the meal was nearly over before Mrs Leader made her entry. She dined with her brother-in-law, and by the time that they came out into the lounge, Jessica had already been persuaded to play. There were but few empty chairs, and Mrs Leader, with a cigarette in her mouth, sat firmly down in Jessica's other throne, and talked. She must have been saying something amusing, for her brother-in-law gave his hoarse laugh and said 'Capital, Edith: tell me another.' Jessica, in consequence, as Edith proceeded to tell another, took her hands off the keys and waited with a wide, martyred smile. That impressive pause failed in its effect for once.

'It was too killing,' said Edith. 'There she was, looking precisely like the witch of Endor, with that silly old man. I never saw a woman——'

Edith became aware of the dead hush.

'I'll tell you the rest afterwards, Toby,' she said, and lit another cigarette.

Jessica proceeded with the slow movement of Beethoven's Fifth Symphony, arranged for the piano, exactly where she had left off. Sighs and 'thank you's' were breathed round the room at the end, and she rose and flitted away to where the padre sat with two elderly ladies.

'A little game of bridge, padre,' she said, 'to get into practice for tomorrow night's tournament? No, I won't be persuaded to play to you again. After Beethoven, what *is* there to play?'

Bridge had already begun, with whispered declarations, during the music, and the padre put his private piece of green baize on the slippery table.

'You played divinely tonight, Miss Jessica,' he said. 'Tum-te-tum . . . wonderful music.'

Jessica looked round for her chair and saw it was occupied. She gave her little sign to the hall-porter, who, rather diffidently, approached Mrs Leader and whispered to her.

'Oh, by all means,' she said. 'But where am I to sit then?'

'Sit at the piano,' said her brother-in-law.

She rose.

'I'll play you that bit out of "Foolish Virgins", if you like,' she said. 'But a rotten piano . . .'

She seated herself at the despised instrument, and without preliminary broke into the scherzo of the ballet-music. She played sketchily and superbly, with quantities of wrong notes but a glorious sense of rhythm. . . . Jessica gave shivering winces at the wrong notes and raised her shrill voice.

'One club, did you say, padre?' she asked. 'Now, partner, I shall be ever so reckless and say one spade. So naughty of me, but—oh, that music—but I shall expect to be scolded. Ah, my poor ears . . . and you go two clubs, padre, after everybody has passed. Well—let me think if I can, but who *can* think with this—let me see, what is the right thing? I don't care: I shall go two spades. . . .'

In fact, this was not so much a declaration at bridge as a declaration of war. Red-mouthed, relentless war.

Jessica lay long awake that night inventing tactics and planning manoeuvres. She had observed with pain that Mrs Leader's performance had given pleasure to the less musical members of the happy family: old Mrs Ward, for instance, who was generally found to be asleep at the end of a Beethoven movement, had been sitting bolt upright in her chair and beating time on the arm of it with her fan; young Mr Innes had applauded loudly at the end, and even the padre had said 'Wonderful execution, surely.' But difficulties never daunted Jessica, they only developed her horse-power, and next day she went out to battle.

She had her bath and massage for her rheumatic wrist early, and returning from the establishment had the pleasure of cutting Mrs Leader dead: the pleasure was only marred by the depressing suspicion that Mrs Leader had not noticed it. She then sat down at the piano and, with her wrist in excellent order, played solidly for an hour. By that time there were many little groups of the happy family scattered about the lawn, and she went from one to the other.

'Good morning, Mrs Ward,' she said. 'What a pianist we have among us now! But how impossible to play bridge, was it not, with those rivers of wrong notes? If Mrs Leader plays tonight during our tournament there will be a marked falling off in our play, I am afraid. . . . Ah, Mr Innes, you and I play bridge together tonight, I think. We really must secure a table away from the piano, or I am sure I shall revoke. . . . Dear padre, going to have a round at putting? I wonder—you have such tact—if you could tell Mrs Leader that we are very serious bridge-ites. Oh, dear me, there she is at it again.'

Brilliant and rollicking strains came from the window of the lounge, and Jessica, protesting that she could not read her paper in that riot, retired to the sunless little garden on the other side of the hotel. But she found it cold there and came back. The lawn was completely empty, and looking into the lounge, she saw that it was full.

People came rather late in to lunch that day; in fact there was hardly anyone there except Jessica and deaf Mrs Antrobus till that meretricious hubbub from the lounge ceased. Directly Jessica had finished she tripped away to the piano and had a real good practice. Long before she finished, her fingers were aching, but she held on till four o'clock, at which hour she usually had tea.

She had hardly left the lounge when Mrs Leader entered it from the garden-door. The piano-stool was rather low, and she sat on all Jessica's volumes of Beethoven.... Jessica hurried back again in order to pretend to write a letter, and then distractedly go away again, with pen, ink-bottle and paper in her hand. She saw her music was not on the piano, where she had left it, and began hunting round the lounge for the melodious volumes. She looked high and low and called the hall-porter to explain her loss. Not till then (apparently) did Mrs Leader guess what she was looking for, and jumped up, saying:

'Oh, I'm afraid I'm sitting on them. So sorry.'

Jessica made the sort of smile which frightens dogs.

'I hope it won't inconvenience you too much if I take them away,' she said. 'So good of you. Many thanks, and apologies for troubling.'

The witheringness of this sarcasm, for which, when goaded, Jessica was famous, had no effect on Mrs Leader. 'Slightly cracked', she thought to herself, and played chromatic scales for a quarter of an hour.

Jessica, trembling with passion but convinced she had inflicted a deep mortal wound, went up to her room. From the window of it she could just see the corner of the lounge where the piano stood. As soon as she observed that the music-stool was unoccupied, she hurried down again and played easy pieces of Mozart till it was time to rest before dinner. Then, by a brilliant inspiration, she locked the piano and hid the key in a brass vase of Benares workmanship. So *that* would insure them against any ear-splitting strummings during the bridge tournament.

Mrs Leader, declining to take part in the bridge tournament, played a rather loud sort of patience with her brother-in-law, and when that was over she attempted to open the piano. But it was locked, the key was missing, and neither hall-porter nor manager nor lift-boy knew anything about it. But so long as there was no Beethoven possible, she did not much care, and, being in want of an ash-try, took the Benares vase of its shelf and found the key. She was an intelligent woman, and instantly guessed how it had got there. She pocketed it therefore, and on her way up to bed hid it on the top of an engraving of 'The Monarch of the Glen' which hung in the corridor.

Jessica tripped into the lounge early next morning, with the intention, now that the bridge tournament had not been interrupted by distracting noises, of restoring the key to its place. But when she held the Benares vase upside down nothing fell into her hand but some burned-out cigarette-ends. This was both disgusting and disquieting, for she felt sure she had put the key there.

She washed her hands, and went off to her breakfast completely puzzled. Then she remembered that there was another vase of Benares ware at the further end of the shelf, and returned to see if it was there. She had to get on the sofa to reach this, and at that moment Mrs Leader entered.

'Good morning,' she said cheerfully. 'Looking for the key of the piano? Have you forgotten where you put it?'

These remarkable words gave Jessica quite a shock, and she had to steady herself against the shelf.

'I beg your pardon?' she said in her iciest tones.

'Pray don't mention it. But it is annoying to have put something carefully away and to forget where you put it. Sunday too: we shan't have any of your delightful Beethoven.'

Jessica dismounted from the sofa.

'I am quite at a loss to understand what you mean,' she said.

'I'm afraid I can't explain myself more clearly,' said Mrs Leader. She broke into a shout of good-natured laughter.

'You put it on the top of the picture of "The Monarch of the Glen"', she said. 'If you'll fetch it, let's sit down and have a duet. But no more hiding, mind!'

DESIRABLE RESIDENCES

HOUSES in Tilling are in much request during the months of August and September by holiday-makers of the quieter sort, who do not want to stay in large hotels on esplanades in places where there are piers, to flock to the shore in brilliant bathing-costumes, to pose for photographers in the certainty of winning prizes as plump sea-nymphs, to dress for dinner and dance afterwards. But families in search of tranquillity combined with agreeable pastimes, find Tilling much to their mind: there is a golf-links, there are illimitable sands and safe bathing: no treacherous currents swirl the swimmer out to sea when the tide is ebbing (indeed, the shore is so flat that the ebb merely leaves him stranded like a star-fish miles away from his clothes): there are stretches of charming country inland for exploratory picnics, and Tilling itself is so full of picturesque corners and crooked chimneys and timbered houses that easels in August render the streets almost impassable.

The higher social circles in this little town are mainly composed of well-to-do maiden ladies and widows, most of whom, owing to the remunerative demand for holiday residences, live in rather larger houses than they otherwise would and recoup themselves by advantageous letting. Thus towards the middle of July a very lively general post takes place.

Those who own the largest houses with gardens, like Miss Elizabeth Mapp, can let them for as much as fifteen guineas a week, and themselves take houses for that period at eight to ten guineas a week, thus collaring the difference and enjoying a change of habitation, which often gives them rich peeps into the private habits of their neighbours. Those who have smaller houses, like Mrs

Plaistow, similarly let them for perhaps eight guineas a week and take something at five: the owners of the latter take cottages, and the cottagers go hop-picking.

Many householders, of course, go away for these months, but those who remain always let their own houses and are content with something smaller. The system seems to resemble that of those thrifty villagers who earned their living by taking in each other's washing, and answers excellently.

Miss Mapp on this morning of early July had received an enquiry from her last year's tenants, as to whether she would let her house to them again on the same terms. They were admirable tenants who brought their own servants, a father who played golf, a mother who wrote letters in the garden, and two daughters with spectacles who steadily sketched their way along the streets of the town.

Miss Mapp instantly made up her mind to do so, and had to settle whether she should take a smaller house herself or go away. If she could get Diva Plaistow's house, she thought she would remain here and take her holiday in the winter. Diva was asking eight guineas a week, including garden-produce. The crop on her apple-trees this year was prodigious, and since garden-produce was included, Miss Mapp supposed she would have the right to fill hampers with what she couldn't eat and take them away at the end of her tenancy.

'I shouldn't have to buy an apple all winter,' thought Miss Mapp. 'And then fifteen guineas a week for eight weeks makes a hundred and twenty guineas, and subtract eight times eight which is sixty-four (I shall try to get it a little cheaper) which leaves—let me see . . .'

She arrived at the sumptuous remainder by tracing figures with the handle of her teaspoon on the table-cloth, and having written to the admirable tenants to say that she would be happy to let her house again at the same price, hurried to the house-agents to make enquiries. She could, of course, have gone to Diva direct, but it would not be pretty to haggle in person with so old a friend. She put on her most genial smile, and was artful.

'Good morning, Mr Hassall,' she said. 'A cousin has asked me to enquire about houses in Tilling for the summer. I think Mrs Plaistow's little house might suit her, but I fancy she wouldn't pay as much as eight guineas a week.'

'Very nice house, ma'am. Very good value,' said Mr Hassall. 'Garden-produce included.'

'Yes, but eight guineas is rather high. But perhaps you would tell Mrs Plaistow that you've had an enquiry offering seven. And what about servants?'

'Mrs Plaistow is thinking of getting another house for the summer, and taking her servants with her.'

Miss Mapp considered this, still smiling.

'I see. Then would you make enquiries, and let me know as soon as possible? I am going home at once. Good morning. What a lovely day!'

This question about servants was, like all Miss Mapp's manœuvres, much to the point. If Diva was leaving servants, her plan was to pick a quarrel with her cook without delay, and give her a month's warning, which would bring her to the beginning of August. But there was no need for that now.

Miss Mapp stepped out of the office into the hot sunshine, and failed to observe Diva, round and red, trundling up the street behind her. But Diva, whose eyes were gimlets, saw Miss Mapp and where she came from, and popping in to see whether there were any enquiries for her house, heard from Mr Hassall that he had just received one, offering seven guineas a week. Such evidence was naturally conclusive, and she had not the smallest doubt that this nameless tenant was Miss Mapp herself.

Mr Hassall allowed that the enquiry had been made by Miss Mapp on behalf of a cousin, and Diva laughed in a shrill and scornful manner. She no more believed in the cousin than she believed in the man in the moon, and it was like Elizabeth—too sadly like her, in fact—to attempt to haggle behind her back. She also drew the inference that Elizabeth had received an offer for her house, and already rolling in prospective riches, wanted to roll a little more.

'Kindly ring Miss Mapp up at once,' she said, 'for I saw her going up the street towards her house, and say that I am asking eight guineas a week, and will not take less. I should like a definite answer at once, and I'll wait.'

The telephone bell saluted Miss Mapp's ears as she entered her own door, and the ultimatum was delivered. It was tiresome to have used the cousinly subterfuge and have got nothing by it, but the difference between even eight guineas a week and fifteen was quite pleasant. So she accepted these terms, and since it would soon be obvious that she was her own cousin, she admitted the fact at once.

Diva was so pleased to have seen through the transparent and abject trick so instantaneously, that, full of self-satisfaction at her own acuteness, she bore poor Elizabeth no grudge whatever. She only sighed to think how like Elizabeth that was, and having thus secured a very decent let, inspected a smaller house belonging to Mrs Tropp which would suit her very well, and obtained it, for the period during which she had let her own, at four guineas a week.

Some fortnight later, Miss Mapp was returning from an afternoon bridge-party at Diva's. She had won every rubber, which was satisfactory, and had caught Diva revoking beyond all chance of wriggling out of it, which made a sort of riches in the mind of much vaster value than that of the actual penalty. But it was annoying only to have been playing those new stakes of fourpence halfpenny a hundred. This singular sum was the result of compromise: the wilder and wealthier ladies of Tilling liked playing for sixpence a hundred, but those of more moderate means stuck out for threepence. Diva who hardly ever won a rubber at all was one of these.

She said she played bridge to amuse herself and not to make money. Miss Mapp had acidly replied, 'That's lucky, darling.' But that was smoothed over, and this compromise had been arrived at. It worked quite well, and was a convenient way of getting rid of coppers if you lost, and the only difficulty was when there happened to be a difference of fifty or a hundred and fifty between the scores.

'If a hundred is fourpence halfpenny,' said Miss Mapp, 'and fifty is half a hundred, which I think you'll grant, fifty is twopence farthing.' . . . So after that, they all brought one or two farthings with them.

Still, even at these new and paltry stakes, Miss Mapp's bag this evening jingled pleasantly as she stepped homewards. But one thing rather troubled her: it was like a thunder-cloud muttering on the horizon of an otherwise sunny sky. For she had heard no more from the admirable tenants: there had just been the enquiry whether she was thinking of letting, and then a silence which by degrees grew ominous.

She wondered whether she had acted with more precipitation than prudence in committing herself to take Diva's house, before she actually let her own, and no sooner had she reached home than she became unpleasantly convinced that she had. The evening post had come in, and there was a letter from That Woman who had written

so many in the garden, to say that a more bracing climate had been
recommended for her husband, and that therefore with many
regrets . . .

It was a staggering moment. Instead of raking in a balance of
seven guineas a week, she would possibly be paying out eight. July
was slipping away, so the pessimistic Mr Hassall reminded her when
she saw him next morning, and he was afraid that most holiday-
makers had already made their arrangements. It would be wise
perhaps to abate the price she was asking.

By the twentieth of July, anybody could have had Miss Mapp's
house for twelve guineas a week: by the twenty-fourth, which
ironically enough happened to be her birthday, for ten. But still
there was no one who had the sense to secure so wonderful a
bargain. It looked, in fact, as if the Nemesis which has an eye to the
violation of economic problems, had awakened to the fact that the
ladies of Tilling took in each other's washing (or rather took each
other's houses) and scored all round.

And Nemesis, by way of being funny, did something further.

On July the thirtieth, Miss Mapp's most desirable residence, with
garden and the enjoyment of garden-produce, could be had,
throughout August and September, for the derisory sum of eight
guineas a week. On that very day two children in the cottage which
Mrs Tropp (Diva's lessor) had taken for herself developed mumps. A
phobia about microbes was Mrs Tropp's most powerful characteristic, and with the prospect of being houseless for two months (for
she would sooner have had mumps straight away than be afraid of
catching them) she came in great distress to Diva, with the offer to
take her own house back again at the increased rental of five guineas
a week.

Besides, she added, to turn two swollen children out into the hop-
fields was tantamount to manslaughter. Upon which, to Mrs Tropp's
pained surprise, Diva burst out into a fit of giggles. When she
recovered, she accepted Mrs Tropp's proposal.

'So right,' she said, 'we couldn't bear to have manslaughter on our
consciences. Oh, dear me, how it hurts to laugh. Poor Elizabeth!'

Diva, still hurting very much, whirled away to Mr Hassall's.

'A cousin of mine,' she said, 'is looking out for a house at Tilling
for August and September. Miss Mapp's, I think, would suit her, but
seven guineas a week, I feel sure, is the utmost she would pay. I

should like a definite answer at once, and I'll wait. Why, if I didn't use exactly those words to you, Mr Hassall, when last you telephoned to Miss Mapp for me! I won't give my name at present—Just an offer.'

Miss Mapp was in the depths of depression that afternoon when the telephone bell summoned her. She had practically determined to stay in her own spacious and comfortable house for the next two months, since it was of quite a different class to Diva's, but the thought of paying out eight guineas a week for a miserable little habitation (in spite of the apple-trees) in which she would never set foot gnawed at her very vitals. Of course with the produce of her own garden and Diva's, she would have any amount of vegetables, and with the entire crop of Diva's apples added to her own cooking-pears (never had there been such a yield) she would do well in the way of fruit for the winter, but at a staggering price. . . .

Then the telephone bell rang and with a sob of relief she accepted the offer it brought her. She hurried to Mr Hassall's to confirm it and sign the lease. When she knew that the applicant was Diva, and divined beyond doubt that Diva's cousin was Diva too, she moistened her lips once or twice, but otherwise showed no loss of self-control.

So for two months these ladies stayed in each other's houses. Mrs Plaistow's letters were addressed to 'Care of Miss Mapp', and Miss Mapp's letters to 'Care of Mrs Plaistow'. Every week Diva received a cheque for one guinea from her tenant (which was the balance due) and another from Mrs Tropp, and immensely enjoyed living in quite the best house in Tilling. She gave several parties there, to all of which she invited Elizabeth who with equal regularity regretfully declined them on the grounds that in the poky little house in which she found herself it was impossible to return hospitalities. . . .

It may be added that on the happy day on which Miss Mapp got back to her own spaciousness, several large hampers of apples were smuggled in through the back door. But Diva had had a similar inspiration, and, scorning concealment, took away with her a hand-cart piled high with cooking-pears.

CRUEL
STORIES

THE PUCE SILK

IT was almost impossible, for physical reasons, that Lucia should ever look annoyed or vexed, because the childlike shape and infantine blue of her eyes, the adorable dimples in her cheeks, and the slightly upward curve of her lips had so irresistible a suggestion of kindly merriment about them. Even on this present occasion, when she was rather seriously put out, no trace of annoyance was present.

She made a little appealing gesture to her husband, throwing her head back and putting forward a supplicating chin.

'Darling, I know it is my fault,' she said, 'and I have told you so. What more can I do? I did ask Aunt Cathie to come here during September, any time that it suited her, and I forgot about it. Well, she has chosen to propose coming tomorrow, on the very day when your first shooting-party assembles, and I shall have absolutely all the wives of the world to look after. I can't see why I should not put her off till the week after, when we shall be quiet again. Besides, I shall be able to devote myself to her next week, which I can't do when the house is full. Aunt Cathie will be hurt if she doesn't see a good deal of me.'

Edgar Brayton was not so fortunate as Lucia: a very distinct shade of annoyance crossed his face at his wife's last remark.

'Oh, Lucia, what is the good of saying that?' he asked. 'You don't want her to come when we are alone in order to be able to look after her. That isn't your reason.'

'It is a very good one,' said she.

'But not yours, so to speak. And your reason, dear—well, I am afraid it is not quite worthy of you.'

Lucia gave a little sigh, not impatient at all, merely misunderstood.

'The reason I have given you seems to me excellent,' she said, 'and I don't know what more you can require of a reason than that. But pray let me hear the unworthy reason which you speak of.'

'As you please. You are a little ashamed of your Aunt Cathie; you don't want her to be in the house with all your other friends. My dear child, she is a lady, and that is all you can require of a woman.'

'Of course she is a lady,' said Lucia, forgetting to disclaim this as being her reason, 'but she is a very odd one.'

'Ah! then that was your reason,' said he.

Lucia got up.

'I don't think we gain anything by arguing,' she said. 'I have told you I think it would be better to put Aunt Cathie off till next week, and you don't agree with me. There is no more to be said. I may remind you that the burden of entertaining her will fall on me, not on you.'

Edgar laughed.

'Nonsense,' he said. 'Aunt Cathie will entertain herself very well. And I have a reason for not putting her off: it is that I know how she will love to see you as the hostess of a big party. It will give her the intensest pleasure. She adores you: she will love to see you shining.'

Lucia was silent a moment, slightly disarmed by this, and making up her mind as to whether she was defeated or not. But there was a quiet finality in her husband's tone which she had learned to recognize, and recognizing that, she resigned. But she resigned with the most charming good temper.

'That is settled then,' she said. 'I am so sorry for interrupting you in the middle of the morning, but I thought I had better see what you thought about this at once.'

He drew her close to him.

'You know you can't interrupt,' he said; 'so write a charming telegram to Aunt Cathie and say how delighted we are.'

Lord Brayton had been quite within the mark when he had said that Aunt Cathie would love to see her niece in the capacity of brilliant hostess. Lucia had lived with her in the extremely interesting but slightly provincial town of Wroxton since her girlhood, and just a year ago had made this wonderful marriage. Not that Aunt Cathie thought it in the least wonderful; it was little surprise to her that Lucia should step up from a semi-detached villa in Wroxton into the full light that beats upon countesses. But it had all been romantic enough: Lucia and he had met at the Dean's garden-party; he had asked leave to call, and before, as Aunt Cathie said, anybody had time to turn round, he had proposed and been accepted. The fairy prince had carried off the fairy princess, and in private the fairy princess's aunt searched *Debrett* to find if aunts of countesses took precedence of anybody. It would be thrilling if they did; but she could find nothing about it.

After that one thought of herself she was entirely absorbed again in the star-showers of Lucia. From the earliest years she had loved the orphan girl with the old-maid love which, barren of natural outlet on husband and children, concentrates itself on girlhood and its possibilities. Even before the wonderful marriage Lucia had shone as no one else had ever shone in the Wroxton firmament, before sailing up into the very zenith of London like this; and Aunt Cathie felt that there would be no intermission in her intimacy with Lucia, for Lord Brayton instantly called her Aunt Cathie, and twice stayed at the semi-detached villa, which now almost lived up to its name, which was Cambridge House.

Aunt Cathie never allowed herself to dwell on disappointments, and so, though Lucia, after her marriage, had not suggested that she should come and live with them, a fate which in more sanguine moments Aunt Cathie had allowed herself to contemplate, she realised that Lucia had now formed new ties, which was quite true, though how true she did not yet know. It was quite natural again that Lucia had not asked her to Scotland in the summer, probably because she knew how fatiguing so long a journey would be, as well as expensive. But now, when she and her husband were to return to Brayton towards the end of September, it was charming of Lucia to ask her to come there any time during that month, even though there was only a week of it left. Aunt Cathie, therefore, proposed herself for three days of that week. It was well to begin gently.

There was a good deal of thought and preparation necessary before this visit, for it was a long time since Aunt Cathie had stayed in what she called a grand house, and on that occasion (the house being that of the Bishop of Wroxton) she felt at home at once. But she was older now, and rather more timid, and most unfortunately she had a rheumatic shoulder, which required a firm hand and liniment. Lucia had often rubbed it for her before; but Lucia, as hostess (and countess) could hardly be expected to do so now, while to be rubbed by a strange house-maid, in case the shoulder proved troublesome, seemed almost indelicate. So Aunt Cathie hit upon a daring scheme, and determined to take her parlour-maid, an elderly and rather effete lady, as maid. That sounded simple; but there was much to be considered, for, to begin with, the parlour-maid must cease to be Jane and become Arbuthnot, which needed practice. Then she had to be instructed as to her social status, and the mysteries of 'the Room' must be made clear to her. Of course this sort of thing was expensive, since Aunt Cathie, if accompanied by Arbuthnot, would have to go second-class (a luxury she never permitted herself), so that Arbuthnot might not feel awkward, if the train was full, in possibly having to travel in the same carriage as her mistress. But Aunt Cathie's quarterly dividends had just come in, and she could manage it quite well.

All this, as has been said, required thought; but when it was settled, it was settled, whereas the question of dress, which had begun to reign in Cathie's thoughts at the beginning of August, in anticipation of this visit, seemed terminable only by the arrival of the event itself. Aunt Cathie was afraid that she was accustomed to think too much about dress; but on this occasion she felt that she must, for Lucia's sake no less than for her own, be richly attired, and she bought several consecutive numbers of *Ladies' Dress* in order to see what people were wearing now, and decide how far she could inexpensively adapt her own wardrobe to this fashionable emergency. Walking dresses, she was glad to see, were very simple, and cut much on the lines of what was known as 'the old speckledy'. But the old speckledy was certainly old, while the new speckledy was, according to the standard required of *Ladies' Dress*, unsuitably florid.

Aunt Cathie shook her head over this.

'The old speckledy would have been just the thing a year or two ago,' she said to Arbuthnot, 'for my walking dress. "Suitable when going out with the shooters" is as like it as a pea—two peas. Perhaps if you ironed it, Jane—Arbuthnot—— Then that, with the blue serge—the one with yellow facings, which Miss Lucia used to like—and the new speckledy, will be sufficient for the day.'

'The evil thereof' did not occur to Jane: the opulence did.

'But you won't take three day-gowns for a visit of three days, miss?' she asked.

'Yes: one for each day. It is always done.'

A tea-gown was the next question, and in about a week, made hideous with heart-burnings and consultation of her pass-book, Aunt Cathie decided against it. With a maid she could easily have tea upstairs when she came in from walking with the shooters in one of the speckledies, and rest till dinner; for no carpentering, however drastic, would make her high evening-gown resemble what 'a lady of title' said was being worn now at tea. But over evening-gowns Aunt Cathie could breathe a sigh, not of resignation, but of high content: she knew she was more than neat in this respect—indeed she was on the far side of gorgeous. Even the second-best was a symphony of grey and lace, and as for the puce-coloured silk which had practically no sleeves——

Jewellery: that was good also. There were the amethysts, necklace, bracelets, and brooch firmly set in pure gold. There were the amber beads which Professor Joblis had pronounced to be very fine for the day; and as a second parure for the evening (or perhaps for a wet day, when there was no walking with the shooters) the double collar of Roman pearls, with the clasp of real ones. The Roman pearls had belonged to her aunt, and were very large and lustrous, and indistinguishable from the much smaller genuine specimens in the clasp. Indeed the Roman pearls had more lustre than the real ones, which were a little stale. But they were but the frame to a superb garnet. The whole garniture was alluded to by Aunt Cathie as 'me pearls'.

Aunt Cathie had arrived, and was resting in her bedroom at Brayton before dressing for dinner. Lucia was coming up for a chat before she went to dress, but Aunt Cathie was glad to be alone for a little and recover from the excitement and strangeness of it all. Three or four motors had been waiting at the little wayside station, with an enormous 'bus for servants, and she had found herself in the middle of a perfect throng of people who all knew each other so intimately that she heard nothing but nicknames or Christian names. There was a Duchess among them, for Aunt Cathie had heard another woman, very rustling in dress, speak to her as 'Your Grace', while everybody else called her Mouse. And the awful certainty dawned on Aunt

Cathie that this resplendent female was Mouse's maid. All the time
Arbuthnot stood like a tall grey monument of despair, until she was
forced into the omnibus, which she entered with the air of one who
took her place on a tumbril. She herself found she was to go in a
motor with three strange men and the Duchess. They had been most
polite and friendly, and Mouse hoped she had plenty of room,
pushed a footstool to her, remarked how early it got dark, and
wondered why the motor crawled like this. To Aunt Cathie it
appeared that they were going the most dangerous pace, and it was a
relief to her when they reached the house.

She knew Lord Brayton, and there was the smooth cheek of
Lucia; but otherwise she felt stranger than a unique specimen in a
menagerie full of ordinary species. The Christian names confused
her most: when she thought that somebody was certainly Tom he
turned out to be Babe. And Lord Mallington was Harry; but there
was another Harry whose surname never reached her memory. Lucia
had been charming, and Aunt Cathie's own name positively rang in
her ears in connection with the people to whom she was introduced;
but beyond her own name, which she knew already, she had grasped
little else. Then at tea she had heard they were going to drive
tomorrow, and putting her courage on the conversational altar, she
had said loudly and distinctly that she would enjoy that very much.
But the drive appeared to be partridges, and even the knowledge
that the old speckledy was similar to that worn by the lady with the
small head when walking with the shooters, did not entirely console
her.

These were her reflections: then Lucia's voice sounded from just
outside her door.

'Yes, Mouse,' she called out, 'at the end of the passage, left-hand
side, you know; but don't go to the right, that's the Babe's room. But
do see he's dressed in time: half-past eight dinner, really half-past
eight, because Edgar dies at twenty minutes to nine if he hasn't
begun. So please be punctual.'

Then there were a few inaudible words, a stifled laugh, and Lucia
asked if she might come in.

Aunt Cathie was sitting with one candle by her, and the room was
otherwise dark. She had thought it strange that there should be only
one candle, but no doubt they would bring a lamp soon. Lucia
entered and spoke.

'But Egyptian darkness,' she said—'pure Egyptian. Dear Aunt Cathie, where are you?'

A sound clicked in the gloom, and half a dozen electric lights flared.

'Or did the light bother you' asked Lucia. 'Shall I turn out some of them?'

'Dearest child,' said Aunt Cathie, 'how stupid you will think me! I never thought of electric light. Beautiful! Two by the dressing-table, one by the bed, and that big one in the ceiling.'

Lucia produced a small cigarette case.

'Oh, how nice to get away from everybody and sit down and be quiet!' she said. 'Yes, don't be shocked, Aunt Cathie, but I occasionally smoke. Only Edgar doesn't like my smoking in my bedroom. Why, I can't imagine, so you see I smoke in yours. Let's see. Tonight Harry takes you down, and he is so amusing——'

Lucia's eye fell on the puce-coloured silk that was laid out on her aunt's bed. She stopped abruptly, and her face came as near to expressing annoyance as it was capable of coming. But she instantly recovered herself.

'Dearest auntie,' she said, 'is that for tonight? It's almost too grand, isn't it? You will find us all in scrubby country frocks, you know.'

A gleam of heavenly triumph came into Aunt Cathie's eyes. The puce-coloured silk was smart: there was no denying it.

'Don't put us all in the shade,' Lucia went on. 'Pray wear something less magnificent.'

Aunt Cathie gave a bubbling sound of pleasure, half laugh, half purr.

'There is another grey dress with lace,' she said.

'Ah! then that would be much better for a higgledy-piggledy party like this,' said Lucia. 'Do wear that. And now I shall sit and talk to you for half an hour, please. Or rather you must talk to me. Tell me about all the gaieties in Wroxton, and how your garden is getting on, and how the servants are. And that nice parlour-maid: Fanny, wasn't it? No, not Fanny—Jane.'

'Arbuthnot is with me,' said Aunt Cathie. 'She is my maid. She was Jane.'

'Then I must see her,' said Lucia. 'Dear me, what funny, dear days those were! Sometime, later on, Aunt Cathie, I shall ask you to let me come and stay the night, and have my old room, and grub in the garden, and play Patience with you after dinner, and go to bed at half-past ten.'

Lucia was quite admirable at this sort of wordy tenderness that meant nothing at all to her but so much to Aunt Cathie, and the 'strangeness' that Cathie had so markedly felt at tea-time was quite evaporated before Lucia found it necessary to 'fly' in order to write a note or two before dressing. As she flew, she cast one more glance at that amazing silk dress laid out on the bed, and congratulated herself on having, so to speak, put her foot on it. She had managed so cleverly too, for she hated to hurt other people's feelings, if she could get her way without doing so.

Aunt Cathie soon rang for her maid, with a feeling of sober elation at the dignity of this. Arbuthnot appeared, from the Room, looking rather dazed.

'Jane,' she said, 'Lady Brayton thinks that perhaps the puce silk is too grand. So I will wear——'

At that critical moment her eye fell upon its shimmering folds, its sleeveless splendours, and the temptation was irresistible. To be grander than duchesses! To make them all feel that they had scrubby country frocks on! Aunt Cathie was mortal: she was a woman.

'I think I will wear the puce silk after all,' she said. 'And here is the key of the case where the amethysts are.'

Cathie spent a perfectly delightful evening. Harry—she tried to read his name off his dinner-card, but her eyes were not what they had once been—had taken her down; he had seemed to enjoy her accounts of the complications of Wroxton Society quite immensely, and she overheard him afterwards repeating the history of the crisis concerning the mayor's niece to the Duchess, who shrieked with laughter. There was no doubt, too, that the puce silk with the amethysts made nothing less than a sensation. She had come down rather late, and conversation ceased altogether for a moment in the drawing-room as she made her shining entrance. But she could not, though conscious of her own splendour, agree with Lucia that the others were scrubby. Lucia, herself, for instance, was dazzling in orange chiffon. Jiminy (whoever she was) wore pink satin with the most lovely lace; and though the Duchess's gown was of the simplest, Aunt Cathie, with her eye acute from recent study of

Ladies' Dress, saw that the simplicity differed somehow from that of the old speckledy. Mouse's pearls, too, were quite as large as Cathie's Roman set, and awe seized her at the thought that these might be real, not Roman. After dinner the others played bridge, a game that Aunt Cathie did not know, though she said she was quite willing to learn if they wanted her to make up a table. But it looked very easy, like dummy whist, which she had played often and often, and she almost repented of her confession that she did not know it, since she was sure she could easily have picked it up. But when she discovered that at the end of two rubbers Lucia had lost nine pounds she felt she had had an escape. How foolish of Lucia! She could not be much of a player.

The only thing, in fact, that marred her evening was Lucia herself. She had seemed almost to avoid Aunt Cathie, and did not even kiss her when she said goodnight. But very likely she was upset at losing so much money; also, perhaps, she was vexed at the appearance of the puce silk. But its wearer had enjoyed it so enormously that she could not regret the risk she took of making the others look scrubby.

She came down punctually to breakfast next morning, and had a very pleasant meal, though she was the only woman there. Harry again was athirst after details of life in Wroxton, but Lord Brayton did not seem amused by them. Then, soon after breakfast, the men went out shooting, and three or four of the women, among whom was Aunt Cathie, drove to take lunch with them at a farmhouse. But afterwards the weather looked a little threatening, and instead of going on with the shooters, in the old speckledy, she walked back alone to the house. It had grown a little chilly, and she took her work, a large worsted comforter, down to the drawing-room to await the return of the others. Nobody, it appeared, was at home; she had the house to herself.

The room where she sat was that in which they had had tea on arrival yesterday. It was of great length, and was broken up by the grand piano in one place and a large screen or two elsewhere. Aunt Cathie established herself behind one of these at the far end of the room close to one of the two fireplaces. She worked for a little, but her needles made but tardy progress, for her walk and this warm room were drowsy influences, and before long she fell asleep.

Half an hour later she awoke: it was growing dark, but the firelight flickered comfortably on the walls. A little distance off on the other

side of the screen which sheltered her she heard a woman's laugh.
Then somebody—a man—spoke.

'Anything diviner than the crisis about the mayor's daughter I
never heard,' he said. 'She told it all over again at lunch. It is really
like a page of Cranford.'

She recognized the voice: it was Harry's.

Then a woman spoke.

'Dear Lucia is quite furious,' she said. 'She told me she had tried
to be diplomatic about the puce silk and the amethysts, and thought
she had succeeded. It's so tiresome when one's diplomacy has no
success. She thought of telling a footman to spill something on it, so
that it could not appear again. How heavenly that people should have
aunts like that!'

She recognized the voice: it was the Duchess's.

'Yes, most heavenly; but it is important that other people should
have them and not oneself,' said Harry.

'Quite so. Harry, I must get the story of the mayor's niece once
more, and I do hope she will wear the famous silk again. It killed
Jiminy's pink quite, quite dead: the pink never moved again. And
Raikes tells me she has the most wonderful maid, who appeared in
the Room last night in white braces and an apron like parlour-maids
in Mr Shaw's plays. And the bridge! Didn't you hear? She said she
would like to learn.'

Aunt Cathie quietly gathered up her work. Just in front of her, on
the other side of the fireplace, was a door leading out into the hall.
She could escape through that, if she could open it noiselessly,
without betraying her presence. She could not be discovered sitting
there, and she could not wait to hear more. She had heard quite
enough.

She went up to her bedroom, and sat there a little, doing nothing,
with her knitting fallen on the floor and her hands trembling.
Sentence after sentence of what she had heard repeated itself in her
brain: they were going to get her to tell the story of the mayor's
niece again; Lucia was furious; it was good that other people should
have such aunts; Lucia had thought of getting something spilt on her
dress. Aunt Cathie did not cry easily, but a couple of small difficult
tears rolled down her cheeks. She had been enjoying herself so
immensely, she had thought she was having such a success; but now
what of the puce silk and the pearls and grey silk yet unseen? It was

all a mistake: Lucia was furious: Lucia was ashamed of her, and the others laughed at her.

But what was to be done? One thing she knew was impossible: she could not meet Lucia and her guests again. Aunt Cathie had her share of courage, but that ordeal was unfaceable; she could not consider the possibility of it. Nor could she even tell Lucia that she must go away: she must somehow go away without Lucia's knowing it, and leave a note for her. And a plan, blurred at first, became clear.

She rang her bell for Arbuthnot—she, too, it seemed, was as unsuitable to the Room as was her mistress upstairs, though the braces and apron were quite new. When it was answered:

'Jane,' she said, 'I find I must get home at once. Please attend very carefully. I am going to put on my things, and I shall walk to the station. It—it will do me good. Then I shall send back a cab for you and the luggage. You will pack at once, and when the cab comes get somebody to put the boxes on it. Whoever it is, give him half a crown, and give the butler five shillings from me. Also give him the note I am going to write, and ask him to take it to Lady Brayton after you have left.'

For a moment Jane's face brightened; then she thought of her mistress.

'But it's raining, miss,' she said.

'I am sorry for that,' said Aunt Cathie.

The note had been rather hard to write; the butler brought it to Lucia about an hour afterwards.

My Dearest Lucia,

I think I made a mistake in coming to see you when you had a big party with you. I am not accustomed to it, and I felt a little strange. They thought me a little strange too, and so you must forgive my rudeness, because I have gone home, since I did not feel that I could do any better if I stopped. Dear Lucia, it was such a pleasure to see you in your beautiful home with all your guests. Pray forgive me, darling, and make some good excuse for me: you are so clever.

Your most affectionate aunt,
Cathie.

THE GODMOTHER

THERE can be no doubt that Mrs Shuttleworth had beautiful bones, without which every middle-aged face that still strives to present the credentials of youth, wears an aspect of effort, while hers even at the age of forty was as effortless as a thrush's song. Owing to this inestimable advantage she was under no necessity of spending her valuable time over it; she washed it, and it was ready to take its proper place in any assembly and to produce its due effect. It did not sag or wrinkle or collapse into bags and pockets or bulge in unbecoming places, or, indeed, play any of those disconcerting tricks which less happily constructed faces indulge in when youth has permanently played truant from them.

Thanks to the exquisite design of its foundation it remained young and shapely and smooth in spite of the grim passage of the trampling years, while her hair, which curled into the most fascinating little golden knots, did so quite of its own accord, whatever anybody said. Anybody did say so, and a few, across whose path she had passed with all the luminous unconsciousness of a comet, were prepared to listen to assertions that she dyed her blue eyes and enamelled her pretty teeth. But they were quite mistaken; her face, which played a useful and highly respectable part in her fortune, was completely genuine.

The hours that she saved from the performance of outside repairs she devoted to thought. Her head was full of plans as a queen-bee is of eggs, but unlike those busy little creatures with stings, she permitted no army of drones in her hive. Her thoughts were all active workers, and brought home every day their portion of thriftily gathered honey which would keep her in comfort when the inevitable winter came on. She was always busy, and brought a great deal of sound second-rate talent to her businesses, which included bridge, a charming little hat-shop, and a steady output of crayon-drawings of her friends, which for some years now had had a considerable vogue.

Her women all looked thoughtful but dewy, and the men all looked like women. Engaged couples used frequently to give Shuttleworth drawings of themselves or of each other to each other, and quite reasonable little cheques to the artist. Like a sensible practitioner she did not produce too many of these, for a glut would have spoiled her market. Often she regretted that she had no time to execute a commission, or was sure that she couldn't do justice to the extraordinary good looks of her proposed sitter; then sometimes she would relent, and the girl would kiss her and say, 'Oh, Evie, how good-natured of you; you are a darling!'

But the hat-shop and the bridge and the studio were by no means the total of her activities; indeed, they would have done little more than pay for the rent of her delicious doll's-house in Mayfair, which was full of beautiful things, and in which so many amusing gatherings took place. She had been left very poorly off by her husband, who had died some four years before, and by her own exertions she had managed to keep herself in comfort and her boy at Eton without ever running into debt or behaving herself unseemly.

She always had charming clothes (and he a proper number of top hats). She lived among extravagant people, kept a motor and shared in all the amusements of the world, and scandal, quite wisely, passed her by, for there was no place for the sole of its foot. All her old friends, of whom she had numbers, said she was a dear, wasn't she? without the nuance of expecting to be disagreed with, and all her new friends, of whom she had many more, were enthusiastic about her.

To the psychological novelist, that was the only dubious point about her—her new friends seemed to be more devoted to her than her old friends. But, after all, many people show the best of themselves on first acquaintance from the very natural desire to produce a good impression as soon as possible. Who shall blame them? The instinct is a perfectly amiable one, and does not argue any insincerity.

Evie was dining one night in May at the Ritz, and arriving, as she always did, very punctually, found that her companion had not been equally considerate, and that she had to wait. She was not in the least annoyed, for she was thinking out a hat; also it was quite pleasant to watch the incoming tide of diners moving through the cage of the revolving door. There were many friends with whom she exchanged a greeting, and this unpunctuality on the part of Margaret Orde would only entail the loss of the first act of *Aida*, which was less hard to bear than her losses since lunch at the bridge-table. They were not very large, but they were annoying, and subconsciously her mind was delving after some of the more substantial and certain means of revenue.

Then by degrees she became less interested in the new arrivals than in a woman sitting alone on a sofa near her, and wondered vaguely at first and then more actively who she was. At a guess her age might be thirty; she had a face irregularly charming, and full of an alert vivacity which was reflected in the quick, sharp movements of her head and hands. She looked this way and that with the staccato motions of a bird, giving a good many sideways glances at Mrs Shuttleworth, who decided that 'thirsty' was the epithet which, singly, best described her alertness. She spoke to no one, no one spoke to her, and she did not seem to be waiting for anybody so much as to be waiting for everybody. Still, there were quantities of women who might any day be seen thirstily waiting at the Ritz who would not have aroused the faintest curiosity in Evie Shuttleworth's mind, and it was not the woman herself who merited interest so much as her decorations.

'Paris' was written large on her gown; Golconda and the deep sea were inscribed on her jewels. She had a collar of superb diamonds round her neck, and a string of pearls that surely ought to have been a recognizable ticket, but Evie felt sure she had never seen those iridescent marbles before. She had an admirable memory for faces also as well as jewels, and yet that vivacious attractive little visage produced not the faintest sense of features seen and forgotten. She felt quite certain that she had seen neither the jewels nor the wearer before, particularly the jewels.

Margaret Orde continued being late, and nobody continued to relieve the solitude of this stranger. Presently Evie was bidden to the telephone, where she learned that her friend had had a collision while on her way here in her motor, which had considerably shaken, though in no way injured her. She had, however, with any amount of regret, gone back home instead of fulfilling her engagement. She hoped that Evie would use the vacant box at the opera.

Evie was profuse in sympathy, and instantly her bees began to gather on the threshold of her well-ordered hive, and to wing their way in search of honey-bearing pleasaunces. In the bureau, close to her telephone-box, was the amiable manager, with whom she often exchanged a pleasant word or two in French, which she spoke quite admirably, and here again her habitual cordiality bore fruit, for he was delighted to answer a few remarks of Evie's which were almost too discreet to be called enquiries. The lady in question—he took a few tiptoe steps down the corridor to confirm her identity—yes, the lady in question was Mrs Jordan, who was occupying a suite in the hotel, pending her installation in her house in Berkeley Square. She had arrived a few days before, and seemed quite alone; no one asked for her, nor she for anyone.

As for her jewels, well, Madame had seen them, and they deserved a closer inspection. Her maid brought them downstairs every night, and they were lodged in the strong-room. Evie walked thoughtfully back to the open waiting-room opposite the revolving door. Her meditation was swiftly concluded, and she proceeded, as was her custom when she had made up her mind, to act on her intentions without delay.

As she passed close to Mrs Jordan she resorted to the crude but effective manœuvre of dropping her fan, and with one of those bird-like quicknesses the other woman bent down to restore it. She got one of Evie's most cordial smiles in thanks.

'How kind of you,' she said, 'and very awkward of me. I drop everything except my friends.'

That was rather happy; it encouraged something more than a mere bow in response. She got a little more than that, a few words very precisely spoken.

'It is nothing,' said Mrs Jordan. 'I hope you have not broken your beautiful fan.'

Evie sat down again on a chair rather nearer to Mrs Jordan than the one she had occupied before.

'I see we are in the same position,' she said, deciding to sink the fact of Margaret's accident. 'Both of us are waiting for somebody who will not come. I have settled to give my friend five minutes more, and then have dinner without her. She must have forgotten about me altogether—you see my friends drop me, though I don't drop them—and though to forget is the best possible excuse, it shows that you didn't look forward to the meeting much. One doesn't forget the things one expects to enjoy. But dining alone is odious, isn't it?'

This was rather a long introductory speech, but Mrs Jordan seemed to enjoy the points of it. She laughed.

'I'm getting used to dining alone,' she said. 'I have dined by myself every evening since I arrived. And I am not like you, waiting for anybody. I am just waiting for eight o'clock.'

Evie Shuttleworth moved a shade nearer again, and put her chin a little upwards, like a child making a request.

'I have such an audacious suggestion,' she said, 'that you will probably send for the manager and demand protection. I was going to propose that we should dine together. It would be a real kindness on your part; I do hate dining alone so much.'

A faint expression of surprise came into Mrs Jordan's face, but she showed no signs of sending for the manager. Ever so little a moment after the surprise came pleasure.

'I'm sure I should be very much pleased,' she said.

Evie was far too clever to force the pace in any way at dinner, or say anything that should evince eagerness as to a rapid ripening of the acquaintanceship, letting their association for the moment be no more than the association at the same table of travellers in a restaurant car. There was an exchange of names, and after that she asked no question whatever, though her sympathetic comments elicited quite a quantity of information. Mrs Jordan, it appeared, had French blood, and was the widow of a Canadian; she had now left the territory of her late husband's fruit-farms—familiar probably to Mrs Shuttleworth on the labels of bottled peaches—to the management of her brother-in-law, and was settling in England. She had expected to be met and companioned in London by a lady who was widely known, she believed, but who had been detained in Paris by illness, and would not arrive for another week. All this came out very naturally, without being fished for, and in turn Evie was allusive about herself and a few friends, whose names she casually introduced, or whom she pointed out at other tables. Among the names was Lady Orde, her expected companion this evening.

'Who gave a dance at Orde House last night?' asked Mrs Jordan.
'Yes, Margaret did indeed,' said Evie. 'Were you there?'

'Gracious me, no! I saw it in the morning paper. I don't know a soul in London.'

Evie had to remonstrate at this.

'Ah, you must allow me to have a soul,' she said. 'Why do you think I have got none? Do I seem so very soulless? And now—what a pity!—I must really be going. I would much sooner stop and talk quietly than go to the opera, but it is already nine. Thank you ever so much for letting us dine together. It was a true kindness to a lonely woman. But I shall not say "goodbye" to you, but only *au revoir*. We shall certainly meet each other again very soon, unless you turn your back on me.'

Though Evie refrained from asking her new friend to the vacant opera-box that night, deciding that it was better to begin gently, and not risk frightening the fish off (she recalled that odious phrase the moment it had formed itself into her mind), she had no intention of wasting any time before establishing further relations. With an insincerity that was as characteristic of her as her beautiful bones, she put it to herself that here was this really charming little woman quite alone for another week in London where, as she so naïvely said, she did not know a soul.

It was clearly, then, the part of any decent Christian to attempt to brighten so dreary a situation. That there might conceivably be other lonely women in London, without the alleviation of the Ritz and rivers of diamonds, had not previously presented itself to Evie's mind nor roused the Christian spirit, but it was never too late to begin. Besides (here sincerity popped out its head like a Jack-in-the-box), who knew what insidious people might get hold of Mrs Jordan and exploit her and batten on her? Accordingly, though Evie went to the opera alone, her hive was buzzing.

Her box, as usual, became a news-agency between the acts, and among the items that tremendously interested her was the appointment of her old friend, Sir John Carrick, to the Embassy at Rome; he was coming to London, so she learned from his sister, in July, and would take up his new appointment two months later. Here indeed was a fresh pasture for her bees, and for half the night, as she lay awake, they worked busily.

Decidedly it was time for her to marry again, unless she intended to remain single for the rest of her days. Hat-shops, crayon-portraits, and such schemes as were now ripening in her head with regard to Mrs Jordan were but temporary occupations, odd jobs, a char-woman's career, and she wanted dignity, a permanent position, a place, too, where the glare of footlights would be on her, and the ring of real events being hammered into shape be in her ears. She frankly acknowledged to herself that she had, so to speak, thrown a fly over John Carrick before, and certainly he had risen to look at it.

There was a streak of the romantic in him which, since he had never met the woman who to him was inimitable and ideal, had caused him never to marry at all, but now, as Ambassador, he would perhaps acquiesce in the fact that most of us live 'in the light of common day', and awake from empty romantic dreams to a very suitable illumination of morning. The disadvantages of being a celibate ambassador were surely overwhelming, and given high romance was not forthcoming, surely he would see the merits of friendship, of mutual liking and esteem. He had no more intimate friend than herself, and she really liked him very much.

Evie woke next morning to a stimulated sense of having, like a conjurer, plates to keep spinning with her deft fingers. A letter of congratulation had to be written to John, carefully worded, which should make his mind suggest situations to itself rather than supply the suggestions ready-made; a crayon-sketch had to be finished; a visit to the hat-shop must be paid, and by the time these duties were performed, it was already nearly half-past one, and she drove to the Ritz in order to make enquiries about the arrival of a friend, who, she knew, could not be in town at all. Very naturally she caught sight of Mrs Jordan.

'Ah, I knew it would only be *au revoir*,' she said. 'I called to see a friend who, of course, is out. Lunching here, are you? My dear, do come and have a very bad lunch with me instead of a very good one. There's an inducement! Another is that my cousin, Lady Arthur Armstrong is coming, who is the greatest bore in the world. Ah, do! That is sweet of you.'

Evie left her alone in her drawing-room for a couple of minutes while she went to tidy herself. She had taken the trouble to bring out from a cupboard a dozen signed photographs of eminent people with coronets and crowns on their frames, which she peppered about the room. She did not know very much about Mrs Jordan yet, and that might produce a good effect.

The preliminary moves might be said to have stopped here, and Evie began being godmother in earnest. She genuinely liked her god-child, her quickness, her adaptability, her really glorious power of enjoyment. Other people liked her, too; she was an instant success, though, with a fidelity positively touching, she never, so to speak, took her hand out of Evie's.

The two in fact slid into their respective roles with the smoothness of well-fitting machinery; each seemed to be made for the co-operation of the other. There on the one hand was Violet Jordan, who, backed by an immense fortune, had come to London with the simple purpose of mingling with the constellations of this spangled firmament, which to her fruit-farming eyes just existed beautifully and sang together like the morning stars. To her this bright environment called 'English Society' was an ethereal circle, a sort of lunar halo, to which, with a golden ladder and plenty of perseverance, you might some day be enabled to climb.

She understood it would be very expensive, but with half the peach-trees of California sunning and ripening for her benefit, there was no difficulty about that. To pour out money, indeed, as Evie soon delightedly perceived, was an actual pleasure to her. But the one essential difficult to procure, probably, was that someone authentically revolving in that starry sky should descend, take you by the hand, and stand sponsor, and lo! to her, while she had yet been but a few hours in London there had appeared, as out of a rosy cloud, this celestial inhabitant at whose girdle swung the keys of all the houses of heaven. Of the other part was Evie, eager to fill her role prodigally and whole-heartedly. With the idea of taking by storm, she had not bothered about surveying and preparing and considering, giving Violet a lunch-party one day, a ticket to the Ladies' Gallery at the House of Commons the next, and a squash on the stairs a couple of days later. She tucked Violet under her arm wherever she went, and as there was nowhere she did not go, Violet spread through London like some delicious influenza which everyone longed to catch.

So urgent a godmother was Evie, that she scarcely sold a hat or made a crayon-sketch all these weeks; as always, she took no partner into her business, and whether Violet was to swim or sink, relied on herself alone. It soon appeared that the result was to be magnificent swimming, for her pupil swept along with great splashes of the divided seas and with hardly time to breathe. She was Evie's sole and peculiar property; nobody else ever ventured to put in a claim for having a part in this splendid launching.

The companion from Paris, so luckily delayed in the first instance, proved to be no more than the widow of some perfectly unknown baronet, whose idea of success was to have a house-boat at Henley, parties at Hurlingham, and stalls at the Handel festival, to sit in the Park on Sunday before lunch, and eventually to be presented at Court. For none of these things had Evie the slightest use, and after the briefest sojourn in town, Lady Jamcock—such was her inconceivable name—returned to Paris with a suitable cheque, astonished to find how London had changed since her day.

All this was easy to one who, like Evie Shuttleworth, really had an immense number of keys that opened doors to her, and she showed her practical ability in obtaining for Violet a lease of Ranworth as a country house, a huge Jacobean residence with six thousand acres of excellent partridge-shooting in Cambridgeshire. It belonged to an odious and impecunious cousin of hers, who for years had been trying to let it at a moderate rent, without any success, for the house was in abominable repair, and the garden a jungle. Over that Evie had effected a very sound stroke of business, for she had offered Lord Ranworth five hundred a year more than he asked about it.

'You're on the make, Evie,' he said, noticing the charming pearls she wore. 'You've got some South African millionaire up your sleeve.'

Evie dimpled into smiles.

'I've nothing of the sort,' she said. 'I say that I can get you five hundred a year more than you asked. My condition is that you take your money and ask no questions nor answer any about the whole business. Just think! You can be at Monte Carlo all the autumn and winter, living in sunshine and cigars.'

The bargain was successfully concluded, and Evie, with the business-like quality which made the hat-shop such a success, took Ranworth off the books of two agents, who furnished details as to price that very afternoon.

Before June was half over, armies of plumbers and decorators had invaded it, gardeners were clipping its lawns, and arrears of wages were paid to astonished lodge-keepers. By the end of July it was ready for the first house-party that would assemble there in the early days of August. To Evie's mind this function, as far as she and her godchild were concerned, rather resembled the rite of confirmation. Violet was now to take on herself the responsibilities which had hitherto been the affair of the sponsor.

The party was quite a small one, but it was potentially momentous not only for the hostess but for Evie also, since it included Sir John Carrick. He had been a few weeks in London, but in the rush and scramble of the season she had seen little of him, and it seemed to her a singularly happy idea of Violet's to secure him for a few days of quiet in the country. He took his hostess into dinner the first night, but Evie easily managed to slide off with him into the moonlit dusk of the terrace afterwards. There were two tables of bridge without them—here were larger stakes.

'It was delightful that you were able to come down,' she said, 'for I saw nothing of you in London. My dear, I was so pleased about Rome; I even presumed on our friendship so far as to feel proud. Tell me all about you and yourself and your plans.'

'You and yourself?' he asked.

'Yes; there is such a real distinction between the two. By "you" I mean the public "you"; by "yourself" I mean the private "you". I want to hear about both.'

'You were always ingenuous,' he said, laughing. 'But you know about the public "me", and as for the private "me", that will tell its own story. Your powers of observation are finer than mine of narrative. What I want you to tell me about is our perfectly charming hostess. I met her two or three times in London, and liked her better every time I saw her. You have a marvellous faculty for finding delightful people. She told me that she was feeling dreadfully lonely and desolate in London, and that you simply picked her up and took her home. She is quite devoted to you. She says she did not know that anyone could be so kind as you showed yourself.'

This was surely pleasant hearing, though not news. For one moment Evie wondered why she did not appreciate it more.

'Violet's a perfect dear,' she said, 'but really she shouldn't bore her guests by singing my praises. I but introduced her to a dozen people, who all liked her.'

'I should think so,' said he. 'There's a freshness, a sort of crystal quality about her that is entirely charming.'

Evie threw her thin shawl, shot with silver thread, rather more closely over her shoulder. Draped closely like that, she looked like some delightful Tanagra statuette, artfully simple.

'Ah, that appeals to you hardened and crabbed diplomatists,' she said. 'And you are quite right. Violet is the most deliciously undiplomatic person I ever saw. Words like intrigue, finesse, tact, are quite missing from her dictionary. She comes to me to ask what they mean.'

Evie found herself giving an ironical touch to this speech without really intending to. But the irony was undoubtedly there.

'I can't agree about her absence of tact,' said he. 'She may not know she has it, like that eminent personage—who was it?—who suddenly discovered that he had been speaking prose all his life. But it was excellent prose, and hers is excellent tact. The very fact that you don't notice it shows its quality.'

Though for the last two months Evie had been delighted to hear the praises of her friend, it struck her now that she had enough of them. Violet was a dear, but like all the rest of the world she had her limitations.

'So magnificent that it is invisible,' she said. 'I must take its magnificence on trust.'

At that moment Violet appeared at the open French window of the drawing-room.

'Are you two not coming to cut in?' she asked. 'A rubber is just over.'

Evie laid her finger-tips on her companion's arm.

'Is that an instance of her tact?' she murmured.

Violet advanced a step.

'Or perhaps you would rather not play?' she added quickly. 'I only came to see.'

Evie smiled brilliantly.

'Bridge is lovely,' she said, 'and your terrace is lovely, and you are lovely. What are we to do? I am greedy for the greatest pleasure. I think we'll send Sir John in to play, and you and I will stroll.'

'You leave me out of your benevolent schemes,' said he.

Violet gave one of those swift, bird-like little nods.

'Ah, you've declared yourself,' she said. 'I shall go in, and leave you and Evie here. But you mustn't monopolize her too much.'

The two strolled on again.

'Such a darling,' said Evie, 'and she does enjoy bridge so enormously. She is far the worst player in London.'

'I consider that very distinguished,' said he.

Three mornings later Evie was finishing a sketch of her friend, putting in with touches of body-colour the highlights on her pearls, when John Carrick came in. The moment she saw him she knew.

'Am I interrupting you?' he asked. 'I hope not. . . . Ah, that is delicious; you were always wonderfully clever. But we are such old friends that I wanted you to be the first to know. Can you imagine what it is to look for one person all your life, and find her when you are forty-seven? Of course you can't!'

A BREATH OF SCANDAL

THE village of Aldwyn, lying at the base of the Dorset Downs, might equally well be described as a small town, for it numbers not less than a thousand inhabitants. It also contains two factories, one of which produces the type of pottery known as Aldwyn ware, which is so rightly designated as 'quaint', and which is so popular with the tourists who visit the place in considerable numbers for the sake of its Norman church and the foundations of its Roman camp. Its other produce is jam, made from the strawberry fields and cherry-orchards which spread far across the surrounding country. Aldwyn is thus a cog-wheel, though a minute one, in the machine of British industry, and may be called a town.

But it has some characteristics of a village as well. Right in the centre of it there is a Green of a dozen acres, where sheep graze round the edges of a good-sized pond. On one side of it stands the Norman church with vicarage of timber and rough-cast adjoining, and round the other three sides are small Georgian houses with gardens fore and aft, each standing widely separate from its neighbours: these give it a rural and spacious air. They are chiefly inhabited by elderly men and their families, who have retired from their professions and spend their days gardening or playing golf on a very decent little nine-hole course by the Roman camp, and their evenings in playing bridge. Some sing in the church-choir on Sunday, having practised their hymns and psalms under the tuition of Mrs Isabel Malden, the wife of the vicar, who plays the organ with a great deal of expression, but not much work on the pedals.

Mrs Malden was a handsome, energetic woman, who had rightly earned the title of the Reverend Mrs Malden. For though she made no direct claim that the world was her parish, her husband's parish was certainly her world. It was her spiritual garden far more than his, and in it she laboured unweariedly with comforts and consolations for those who ailed in body or soul, and, no less, with a copious supply of weed-killer, for the extirpation of evil was quite as important as the cultivation of virtue. She never parleyed with weeds, but was drastic in her dealings with them. She strongly encouraged ill-treated wives to be separated from their husbands, even when they would have much preferred to take the thick with the thin, and muddle along somehow; and she got them situations where they could earn their bread in tranquillity and self-respect. This was useful in times when domestic servants were so hard to obtain, and most of the houses round the Green had a parlour-maid or a house-maid or a cook whom Mrs Malden had torn from her home. But she was fair-minded, and was equally insistent that, when the wife was the erring partner, through drink or other frailty, her husband should get rid of her. So bad for the children.

Indeed, the entire management of the parish was in her capable hands, and since her activities were abetted by a young and vigorous curate, there was really very little pastoral work left for Cyril Malden to do. When he had written his sermon for Sunday morning, his time was his own, and he could devote it without qualms of conscience to the culture of roses and to his definitive edition of the Idylls of Theocritus. While classical tutor at King's College, Cambridge, he had produced an admirable translation of his favourite author into English rhymed verse, and now, having been presented to a college living, he could, owing to his fortunate choice of a wife, continue, with but little interruption, his life's work. He was a scholar of such thoroughness that nothing short of what he deemed the last word would satisfy him, and the typewritten chapters of his book were continually expanding with amplified notes, and being typed afresh. He was a tall, shambling figure of a man, short-sighted, dusty and absent-minded, with a powerful weakness for good food, which, whenever he had the chance, he ate in small quantities with large enjoyment. At home the chance was not often granted him, for the cook at the vicarage was usually a woman whom his wife had torn from her husband, and whose education was lamentable.

His absent-mindedness occasionally caused embarrassment. Not long ago his wife had been dining with a neighbouring squire, and sitting opposite her, that familiar face had caused him to forget that it was not the vicarage table that lay between them. He put down his spoon after one small mouthful of soup, and said in his quiet, distinct voice: 'My dear, I am afraid we must consider your new cook as one of our failures. . . .'

Mrs Malden, coming home today after a long afternoon of district-visiting, noticed that a furniture van was drawn up opposite the door of the house on the Green next to the vicarage, which had so long stood untenanted. A large box, labelled 'books', was at the moment being carried in, and her sight, unlike her husband's, being peculiarly sharp, she could easily read that it bore the name of Trevor Miles. A big woman, not young, but of exceedingly agreeable appearance, whom Mrs Malden's equally sharp perceptions guessed to be the cook or housekeeper, was standing in the doorway, and gave some directions, in a voice that confirmed this conjecture, to the men who were carrying the box. She then disappeared into the house, and two minutes later Isabel was giving her husband his tea.

'"The Lilacs" has been taken at last,' she said. 'Vans are unloading there.'

He came back from a long distance. 'Indeed, my dear,' he said; 'I had observed nothing. Let us hope for the best. Let us pray to be delivered from gramophones. Any indications?'

'A capable-looking cook, and the name of the tenant, Trevor Miles.'

His cup rattled in its saucer.

'What?' he cried, 'Trevor Miles? Surely there can't be two!'

'That I couldn't tell you,' she said. 'I can only vouch for one. But what about him, if he's the right one?'

'Merely that out of all the world he is the man whom I should most like to know. His book, just out, on the Alexandrian age is the most masterly piece of historical criticism. Let us lose no time in being neighbourly. Is there a Mrs?'

'How can I tell?'

'True. But to think of Trevor Miles coming to live here! I will stroll across there and leave my card. If I see him, I will say how glad we shall be if we can be of any use while he is settling in. Meals: that sort of thing, such as they are. No doubt I can ascertain if he is married, and then you will call, too.'

'I shouldn't go just now, if I were you,' she said. 'They'll be all at sixes and sevens. They won't want to see anyone.'

'I fancy I will just stroll across,' said the Vicar.

The newcomer quickly settled in, and was much pleased to find next door a scholar of such repute. He was a small, dapper man, ruddy of face, aged perhaps forty, with a charming manner. He punctiliously returned the cards the Green left on him; he had come down here, he said, in order to do his work. Town-life did not suit him. He played golf quietly and badly, he played bridge quietly and well, and he got into Mrs Malden's good graces at once by engaging, on her recommendation, an aged charwoman to come in early every morning and work for half a day. Otherwise his household consisted only of the woman whom Mrs Malden had instantly divined to be his cook.

There, indeed, was a treasure. Mrs Grainger valeted him, she cooked his meals and served them herself, and those meals, as the Vicar soon discovered, were food for the gods, and of that simple perfection which is only obtained by the great artist. Every morning Mrs Grainger went to do her own marketing, and butcher and greengrocer speedily respected her for her knowledge of 'cuts' and her insight into the heart of a lettuce. Speckless was the house, when she had repaired the shortcomings of Mrs Malden's charwoman, and bright the silver and fresh the flowers on her master's table. Then in the evening, when her ministries were done, she would bring her typing-machine into his study, and he dictated to her what he had written that day. Sometimes there was little, but sometimes, if his work had prospered, the light burned late in his study on the ground-floor overlooking the Green. On these hot, still July nights when windows were open, the clicking of the typewriter could be heard from the vicarage, which so closely adjoined Trevor Miles's house, and the Vicar knew that the learned and diverting essay on Cleopatra, which was now occupying the man who had so rapidly become a friend, was progressing well.

But apart from his historical researches, which interested no one in Aldwyn except the Vicar, the man himself, although pleasant and an agreeable addition to social life, was disappointing with regard to larger aims. Mrs Admiral Woods, for instance, tentatively laid some maternal snares for him with the hope of his marrying her mature but quite well-favoured daughter, and Eugenie was nothing loathe. But the conspiracy was unsuccessful. She was anxious to learn the elements of that admirable stalking-horse for maidenly approach, Contract Bridge, and as he had admitted that it was impossible to learn without dealing out cards and analysing distributions, he found himself committed to tutorship. Eugenie went to tea with him twice, and sat alone with him for a couple of hours, but even her imaginative mind could not fancy that he felt any interest in her apart from the hands he dealt her. He was a charming and courteous tutor, and he felt sure she would soon become an excellent player. Mrs Josephine Draper, the young and attractive widow of a local solicitor, had no better luck. These projects, it must be understood, were of no solid or serious sort; they formed the subject for banter, as Mrs Woods and Josephine Draper sauntered round the Green, and gently chaffed each other on the failure of their matrimonial schemes.

'Just a don, I'm afraid,' said the sprightly Josephine. 'And there's the end of that. But, somehow, it's unnatural that a man in the prime of life shouldn't want a nice woman like Eugenie or me to look after his house for him. But there it is, we must try somewhere else!'

'That Mrs Grainger of his is a remarkable woman,' said Amy Woods. 'Such a worker! I went to lunch with him the other day, and she opened the door to me, and then she served the lunch she had cooked. The house is as bright as a new pin, too. Being too comfortable makes a man celibate: their higher aspirations get drowsy. And she does his typewriting as well, I'm told.'

'Yes, and what a cook! I was dining there a few days ago with the Vicar and the Reverend Isabel. A most delicious chicken-cream—I would have sworn it was chicken. So I said to her: "What a good chicken-cream," for they like little compliments; and what do you think she answered? "Four-legged chicken, ma'am. Bunny!" I felt squashed.'

'Your own fault, dear,' said Amy. 'You should have been able to tell chicken from rabbit, and you deserved it. A handsome woman, too, and how capable! I saw her painting his garden-fence last week; no professional painter could have done it better. And fancy a cook doing that! Why, if I suggested anything of the sort to mine, she would instantly give notice. In fact, I think I shall suggest it. She's one of Isabel's battered lambs. And Mrs Grainger has got a really fine contralto voice. Isabel wants her to sing in the choir, but she said she was too busy on Sunday morning.'

'That's interesting,' said Josephine. 'I didn't know she sang.

Perhaps it explains something that puzzled me. The other night when I was coming back from your house, the window of Mr Miles's study was open, and I heard a woman singing. I couldn't think of anyone here who had such a voice, and it *was* contralto. I confess quite frankly that I tried to look in and see who it was, but the curtain was half-drawn and it hid the piano.'

'I'll tell you something else,' said Amy. 'Eugenie saw him one night when the curtain wasn't drawn. He was sitting at the card-table in the window of his study playing cards with a woman. But she had her back to the window, and Eugenie couldn't see for certain who she was. But she was pretty sure.'

'Rather dreadful,' said Josephine.

'I don't really see why. If a bachelor has a cook who sings well, why shouldn't she sing to him? Or if she plays cards, why shouldn't she play cards with him? Where's the harm?'

'Of course, there isn't any,' said Josephine hastily. 'I meant only just as a matter of taste. There's no reason at all why he shouldn't play cards with her, if he happens to like playing with his cook. It would make me feel uncomfortable myself, but that's neither here nor there. That's what I meant by saying it was rather dreadful, so don't misunderstand me.'

'Yes, dear; I quite see,' said Amy.

But in spite of these disclaimers that either of them harboured any dubious surmise, little uncontrollable thoughts and conjectures crawled about the mind of each of them, like wandering sparks in the ashes of some material that appears to be burnt out.

The active interest in Trevor Miles himself began to shift its direct focus, but it shifted ever so slightly, and now, though nothing was said, Mrs Grainger became the centre of it, he lying just outside. Of course, she was not a lady, and could not, therefore, be considered a lady housekeeper: that was clear from a hundred little indications. Her hands, her movements, her voice, all contributed minute, indefinable items that together made a manifest total. She was just a most admirable cook, who valeted her master excellently, who waited at table like a trained parlour-maid, who painted garden-fences as well as a professional and had a fine contralto voice. Very good-looking, too, rather large-featured, but with big, brown eyes and of a really beautiful, unaided complexion, clear and dark. And how she watched and understood her master! He had but to look up at her without a word, and instantly her eyes scouted this way and that, and she saw what was wanted: the Admiral's glass was empty, or Mrs Malden had dropped her napkin, or Josephine was scraping up the very last morsel of her chocolate-pudding and should be offered some more. Or Amy Woods happened to mention to her neighbour that she had not played bridge for the last three nights, and then Trevor Miles would just nod at Mrs Grainger without interrupting what he was saying, and when they went into his study afterwards, where they sat after dinner, the card-table would be already opened, with clean markers and newly sharpened pencils.

An aloof woman. She kept completely to herself, and had nothing to say to the servants of other houses. The amiable Mrs Jenkins, who had been seven years with Josephine Draper, tried to get into friendly relations with her as they did their marketing, and asked her to step round any afternoon and have a cup of tea; but Mrs Grainger said she was afraid she would never be able to find time. The Admiral's manservant brought a note round himself, and made pleasant conversation, but Mrs Grainger stood firmly in the frame of the back-door, answering him shortly and politely, but not asking him to come in. She wanted to know nobody: her duties were enough for her needs.

Then she fell quite out of grace with Mrs Malden, for, in addition to refusing to sing in the church choir, she told the tottering charwoman that she would not be wanted again after the end of the week. Mrs Malden heard of this, and, since the charwoman had been of her providing, and it was a parochial matter, she came to see Trevor Miles about it. He knew nothing; Mrs Grainger, who managed all matters of domestic service for him, had not consulted him, but nothing would be easier than to hear what she had to say. She came in, thinking he had rung for tea, with his tray. Her explanation was lucidity itself, and she made it to him, quite disregarding Mrs Malden.

'She was no good, sir,' she said. 'I put up with her for a month to see if I could teach her something, but she's past her work. Day after day I had to clear up after she had gone. There's another woman coming in on Monday who, I think, will suit me.'

'Who is that, Mrs Grainger?' asked Mrs Malden.

'A woman whom I engaged, ma'am,' said Mrs Grainger very respectfully. 'Is there anything more, sir? I'll bring another cup.'

There was plenty more next Monday, for Mrs Malden, returning from her morning's visitings, saw, coming out of her neighbour's gate and ironically saluting her, a great flashy woman of odious character, whom her husband, at Mrs Malden's instigation, had lately got rid of. A very bad case indeed: drink and probably worse. Isabel was acutely and genuinely distressed.

'And as you are going to see Mr Miles this afternoon,' she said to her husband, 'I do wish you would let him know what sort of a woman is working for him. I am sure that he can have no idea of it. Tactfully, of course, before you begin that revision of your last chapter. Quite easy for you, as man to man.'

He blinked and fumbled and tried to refuse point-blank. 'I don't like interfering with another man's private affairs,' he said.

'My dear, he leaves all that sort of thing to Mrs Grainger,' she said rather hastily. 'It was quite clear that Mrs Grainger had engaged the woman: he ought to be told.'

Cyril saw his chance. 'Then as Mrs Grainger is responsible,' he said, 'wouldn't it be easier still for you to talk to her, as woman to woman?'

'No. Ultimately, Mr Miles is responsible for his household. Tell him from me that I will find him someone else far more suitable and quite as efficient. He ought not to have that woman in the house, and, after all, dear, you *are* the Vicar.'

That was hardly a fair statement of the case: its superficial truth did not condone its essential falsity. But the titular Vicar promised to do as she asked, and feeling as if he was talking scandal, he hinted to Trevor Miles that his charwoman was not all she should be, and conveyed his wife's promise that she could supply a worker of equal efficiency and vastly superior morals. Miles said it was very kind of her, but evaded any promise to take advantage of her benevolence. Then, in turn, he made an abrupt request. He had long neglected to make his will and thought he ought to do so. Would the Vicar consent to be one of the two executors named therein? His lawyer was the other . . . He was very much obliged. Then the two scholars, pleased to be rid of such dull subjects, happily lost themselves in the vexed question of the date of Theocritus's famous seventeenth idyll. Was it, indeed, by Theocritus at all? Walter Headlam had some very strong arguments against it. The minutes flew by, and Miles had little difficulty in persuading his friend to stop to dinner . . .

The Vicar's intervention had evidently not succeeded, for throughout the next week the undesirable woman could be seen going to and fro from the house. Mrs Malden then spoke to Mrs Grainger about her, forcibly and directly, but, she hoped, persuasively. Mrs Grainger waited till she had completed her black picture, and then answered most respectfully.

'I'm sure I'm much obliged to you, ma'am,' she said, 'but I find the woman very clean and industrious, and she does her work well. She has always been perfectly sober, and I haven't missed anything. Mr Miles will tell me if he wants me to make a change.'

Faintly but definitely after that, Isabel Malden began to wonder what sort of a woman, behind all her industrious efficiency, Mrs Grainger really was. It seemed impossible that a right-minded woman would not have seen the force of what had been told her. A strangely managed household, too (though, of course, that was nothing of her business), in which the cook chooses her underlings and plays cards with her master. She tried to prevent her thoughts from formulating a more specific conjecture than that, and she succeeded in preventing her tongue from so doing. But meanwhile, the unspoken surmise was taking root underground, though no growth of it yet appeared on the surface.

A cold, rainy autumn set in. Waterlogged south-westerly gales discharged their soaking burden, and in the windless spells that followed, the mists rose thick from the Green and shrouded everything with clinging moisture. After a month of it Trevor Miles told the Vicar that he was off to Bournemouth for a week or two in search of less dispiriting weather. He gave him his address at the Dorsetshire Hotel, for there was a further instalment of the Vicar's work on Theocritus which he had promised to look through as soon as it was finished. Off he went, and then it soon began to be noticed that there was no sign of Mrs Grainger at her morning marketings, and the charwoman no longer came only for a few hours in the morning, but lived in the house. She was seen drawing curtains and putting up shutters: evidently she was acting as caretaker. Perhaps Mrs Grainger had gone for a holiday while her master was away.

But there was another possibility, and Mrs Malden, though speaking to nobody of that which grew rank in her mind, felt it was her duty to look into it. She had a most reasonable opportunity of doing so, for a sister of hers lived at Bournemouth, and she always paid her a visit in the course of the autumn. She took with her the new section of her husband's work, and the day after she arrived she went to leave it at the Dorsetshire Hotel. She had a cup of tea, listening to the gay little band that played in the winter-garden, and on her way out left the package at the bureau, and glanced at the visitors' book. Mr and Mrs Trevor Miles were staying in the hotel.

The bureau, at the counter of which she was standing, was close to the door, and at the moment the four-spoked circular cage of entry began to revolve, and the two came in. She turned a little aside and thought she had not been seen by them: they passed not far from her and went into the winter-garden she had just quitted.

Before Mr Trevor Miles and his housekeeper returned from Bournemouth, Aldwyn had made up its mind. No reasonable man or woman could avoid drawing the inference, and it was evident that intercourse must cease, for who, apart from moral considerations, could accept the hospitalities of the house and be waited upon by Mrs Grainger without an embarrassment that would ruin all ease? The Vicar acknowledged the valuable suggestions that Trevor Miles had made on the last batch of manuscript he had sent him, and sighed to think that he could no longer hold friendly council with that subtle and erudite mind. How much keenness and zest he had derived from their discussions! It would be a heavy business to start working again, without the light of that kindling spark.

He finished his letter of thanks, and took up the daily paper, idly scanning the front page before he opened it. Among the deaths was that of Trevor Miles at Bournemouth, the result of an accident. And there was an obituary notice of the most appreciative sort. Next day there arrived for him from the dead man's lawyer a notification of his death, and a copy of his will. It was very short: Trevor Miles had left everything of which he was possessed to his wife, Susan Grainger, known as Mrs Grainger, of 'The Lilacs', Aldwyn.

A week afterwards the widow returned; she asked to be allowed to come and see the Vicar.

'I should like to say a few words to you, sir,' she said. 'For I want to put my husband right with you, and just tell you how it happened.

I was housekeeper to Mr Miles for six months before he came to settle here, and—and we fell in love with each other, sir. He wanted to marry me at once, but I refused. I told him I should leave him.'

'But why?' asked the Vicar.

'Because I was so fond of him, and I knew he would be marrying beneath him. But it was hard work to say no, and presently I told him that I would marry him on condition that no one knew, and that I went on working for him.'

She paused a moment.

'It wouldn't have done, sir,' she continued, 'though it was a long time before I could persuade him of that. I'm not a lady: I couldn't mix with people in his rank of life without being awkward myself and making them the same. And they'd have been saying, "Poor man, to have married one like her." I should have been a millstone, you may say, and I wasn't going to be a drag on him. All I could promise was that if we had a child I would do as he wished. I'm sure I was in the right of it. He'd often have felt half-ashamed of me, however fond he was of me. As it was, he was never that.'

Her eyes suddenly swam with tears.

'He died at once, thank God,' she said. 'One of those great trams went over him. I ran to him, and he lay for a second or two looking at me, as I bent over him, and he smiled at me. Then it was over. Excuse me, sir: I'll get hold of myself again.'

Presently she stood up.

'That's all, sir,' she said, 'and thank you for letting me tell you.'

'But what are you going to do?' he asked.

'Get another situation as soon as I can, sir.'

'But—but, Mrs Miles, you're very well off.'

'Yes, sir, but more than ever now I couldn't get on at all or face it without working. There's an emptiness, oh, such an emptiness! All that I cared for has gone, and I should go crazy if I had nothing to do but sit and think of it. I've advertised as Mrs Grainger, and I shall soon get a new place. I'll be going now, sir, and thank you.'

'But you'll see my wife, won't you? She would like to see you.'

She hesitated.

'Would you mind telling me, sir, what she made of it, when she looked at the visitors' book in the Dorsetshire Hotel? I saw her, though I don't think she knew. Did she mention anything to you?'

Again there was a pause.

'I understand,' she said. 'So I think it's best not to see her. But very natural, I'm sure, that she should think that.'

TO ACCOUNT RENDERED

EVEN in those first intoxicating days of the success which had been so grimly fought for and had tarried so long in its coming, Hugh Ranworth knew that somewhere in the sober background of his mind the question of what it had really cost him kept whispering itself. It was not that he grudged a single hour of that long tale of laborious days, and the series of constant disappointments had left him unembittered. Those were amply compensated for in the reward and the attainment which at length were his; and whatever had been the cost in that species of payment, it was cheap.

But, as some restless and insistent voice reiterated, there was another sort of bill which must have been sent in, the items of which he would some time have to examine. Whatever it was, he had certainly paid it, for in the back of the soul those years of fruitless industry and grim determination could not but be represented by substantial credits and expenditures. But just now he had no desire to peruse it and see how his psychical affairs stood. Whatever that pass-book should reveal, the supreme matter at this moment was the knowledge that he had attained. He had justified his belief in himself; he had got there with a vengeance.

For the last ten years his industry had never abated, nor had his faith in himself burned dim. He had made up his mind that he would succeed as a playwright, and these long years of failure to get any play of his put behind the footlights had annealed rather than weakened his determination. Drama after drama, as each was finished he had sent out on its weary round of rejections, while he busied himself with the next. Like bread cast upon the waters each had returned to him after many days from the actors and managers, who politely thanked him for allowing them to read his most interesting play, but who did not 'see their way' to produce it. Most of these refusals were accompanied by the hope that Mr Ranworth would permit the writer to see any future work of his, and Mr Ranworth had granted them that favour with admirable frequency. Sometimes he seemed to have got nearer to the realization of his ambition than this, for once or twice he was asked to rewrite, on certain lines and for reasons stated, this scene or that, and submitted the manuscript to the managerial eye again. Sometimes even he had been invited to come and talk over certain alterations which might perhaps render his charming play acceptable; these generally consisted in amplifying and strengthening the role of the principal character. 'You must increase the sympathy for me at this point, Mr Ranworth,' was the tune of these suggestions. 'They will want something stronger from me at this point; you do not give me an effective exit.'. . . . Diligently and obediently he would work on these new lines, conscious usually in himself that he was spoiling his play, but eager to do anything for the sake of getting his chance, and bringing his work before the public. Whether or no he was falsifying his talent by such concessions he did not trouble to consider. He told himself, so far as he thought of it at all, that when once he had made his footing he would write plays that satisfied himself as well as the vanity of the insatiable stars. . . . But the conclusion of all these tinkerings and emendations, these blowings of iridescent soap-bubble-soliloquies for the principal actor were always the same. Back came the play on his hands with further vague compliments, and he must set to work again.

He had scribbled and studied unremittingly through these lean, unrewarded years. To analyse the secret of success, to see what made it, and himself to practice it was his unwearied aim. He read and reread plays bad and good which had succeeded, he framed plot after plot which conformed in construction to the methods of his models, thus perfecting his perception of what was dramatically 'telling'. He studied the values of climax and suspense, and, ah! how he studied what the public expected of its favourite actors, and what its favourite actors expected of their ministering dramatists! He learned to write brisk, incisive dialogue, with phrases flashing out like the play of rapiers across the footlights, conveying some subtle, sub-acid, salacious half-truth, with little tinklings of cynical cymbals

which pleasantly challenged the attention. Not less diligently did he
learn the art of peptonizing his thoughts for the audience; it would
never do, so his assiduous study of the masterpieces of popular fancy
convinced him, to make them think. The average well-fed man and
woman towards the close of day does not want to think, and with a
discerning wisdom he learned to do the audience's thinking for it,
and delude it into the belief that it was very clever itself. It wanted
peptonized thought, easy of assimilation. . . .

He consciously began to despise the audience which some day, he
was determined, should discover and acclaim him, and not less did
he despise the popular favourites who were to interpret him. To his
mind they cared nothing for the play beyond that it should give them
an opportunity to hold the stage in the stunt that was expected of
them. One must exhibit himself clad in pyjamas, another in
magnanimity, a third in smart epigrams, tight skirts and loose morals,
a fourth must swim in billows of sentimentalism.

And then, a fortnight ago only, he got there with his new play,
which he thoroughly despised, too. From the first raising of the
curtain till its final fall after an unprecedented number of calls, its
success was complete. Those whose favour he had sought so long
and vainly, now sued to him; and with a chuckling gusto he brought
out from his store of rejected pieces the plays they had refused
before. But now they rose at them like feeding trout at the may-fly,
and two of them were going into rehearsal immediately.

Hugh Ranworth had already stopped in London after that evening of
triumph longer than he had intended. He had promised to go down
a week ago to Blakeney, one of those mellow little red-roofed villages
on the coast of Norfolk, where his mother was expecting him. But
these arrangements with regard to future productions had detained
him, and now it was with a puzzling mixture of eagerness and
reluctance that he set out. He, her only child, was genuinely fond of
her; he knew well that he owed it to her generosity that he had been
able to continue his unremunerative work for so long, confident of
ultimate success. She, as confident as himself about that, had eagerly
economized for his sake, and a certain old daydream of theirs ought
now to be translated into actual fact. Long ago they had settled that
when the grand success came, she should come up to London and
keep house for him there, until he had a wife and a house of his

own; and though Hugh was far from quarrelling with the prospect now that it had moved from the infinite distance into the immediate foreground, he was aware that he did not look forward to it with any eagerness. There was no question at all in his mind of reconsidering it, or by any tepidity of welcome conveying to her that he did not eagerly desire it. Of course, the matter should be arranged at once; she must indeed come up to town with him when he returned, and help him to look for a suitable flat. . . . A furnished flat must be found at once: that would be the best plan. There was something not so permanent about it. . . . Wondering at his indifference, he suddenly perceived that nothing in the prospect of the future gave him any rapture of anticipation, and he told himself, with something of relief, that this apathy was a most natural reaction. It would pass; he would recapture eagerness again.

It was the close of a hot day when he arrived at the dignified little Georgian house facing the quay at Blakeney, and a certain thrill of home as he saw the familiar figure waving hands of welcome at the door made some dim vibration in his heart. What rapture of return there had been from term-time at school to holidays on some such summer evening as this, with weeks of enchanted days of sailing and bathing stretching on ahead, and how as the carriage slowed to take the difficult curve from the quay into the small garden in front of the house, had he often leaped out and run up the short drive. Very vividly now flashed that memory into his mind and he marvelled at the vanished emotion which had then inspired him. But it was jolly, all the same, to get home.

Certainly no such fading of enthusiasm had touched his mother. It was she who ran out to him now.

'Ah, welcome, my dear,' she said. 'Oh, I do call this nice! Hughie, I ought to have hired a brass band to play "See the conquering hero comes!" and I would, too, if I hadn't wanted to have you all to myself. The play, oh, my dear, the play! I knew all along that a day like this would come, and now it's here; and, my darling, the joy of it! I want to hear, at immense length, everything about it all. Never mind your luggage or anything else. Come straight out into the garden with me and have tea and begin!'

There was a long narrative with many interruptions. Hugh had to sketch for her the outline of the play, for though he had brought down a copy of that for her she had to have the first taste of that evening at once, and she must understand, just vaguely, how the play went.

'And you'll have to read it aloud, every word of it,' she said. 'Ah, do; read it me after dinner tonight; that will be delicious! But now just tell it me shortly and describe how it went off.'

The interruptions began to worry him a little. She had to know precisely what Miss Julia Wavely, who took the leading woman's part, was like, and what sort of voice Mr Ames had. All the things so intimately part of his life, matters of consciousness rather than of perception, and hard to put into words, had to be explained and unravelled; it had not occurred to him how completely his mother was a stranger to all that existence meant to him. . . . She got puzzled over the development of the plot; it had to be elucidated again with more particularity. . . .

'I'll read it you this evening,' he said at length. 'You will realize it all quite easily then. And now about yourself——'

'My dear, there isn't any "myself",' she said. 'There's only been a great longing to see you and hear all about it. I can't get to sleep at night, and I don't want to, for thinking of it all. I've enjoyed lying awake and picturing your coming. My dear, how I thank God for it all. And haven't you deserved it too, Hughie? Haven't you just? How your father would have delighted in it! You get your love of the drama from him, you know, though I should like to think you got it from me. . . . And when you came on at the end, and bowed! Oh, what a lovely moment! It was wicked of you to tell me I mustn't come up for the first night, for fear it should be a failure, and I should see you disgraced. Disgraced, indeed! And when, oh when, am I going to see it?'

He envied her the keenness of her happiness, and at the same time the expression of it bored him. He could not rise to its level; he could not enjoy her happiness as she enjoyed his success. But he did his best to stifle the numbing sense of his own incompetence and to grow warm with her glow. . . .

Her last question fairly introduced the mention of the old lovely bargain between them, for which the time was now ripe.

'You're coming to see it when I go back to town, mother,' he said. 'You're coming up with me, and we'll go together on our first evening. And do you know what you're to do the day after?'

'No, dear, what's that?' she said. 'Something heavenly, I know.'

'Something you've got to look for with me,' he hinted.

He saw that she understood, but she did not at once reply. Instead those kind grey eyes grew dim, belying the extraordinary happiness of her smile. She moved a little nearer to him, laying her hand on his.

'Oh, Hughie,' she said. 'I don't pretend not to understand what you mean. It's that we've talked of so often. But, my dear, are you sure you want to be bothered with your doddering old mother always at your elbow? I shall understand so well if you don't. But I bless you, my dear, and thank you for thinking of it. And here am I like an old goose, crying——'

'But of course I want you,' he said. He felt he needed to assure himself of that. 'You're not to be allowed to smash up our joint plans. I won't have it.'

'But listen to me, my darling,' she said. 'We made that plan when you were quite a boy, and needed someone to look after you. But you're a man now; you're thirty-two, Hughie, and you're accustomed to lead your own life. Besides, dear, I hope that soon there'll be some other woman who will look after you. She would turn me out quick enough, and quite right too, if I was such a donkey as to dream of stopping. But even until then, dear—and I hope that day may be ever so soon—it would be so utterly natural if you felt that you needed independence. . . . Shouldn't I understand that, if you told me so, without a single back thought!'

He had done it now. He was not the least sorry, nor wished his words unspoken. Only . . . he wondered how it was that he felt so emotionless.

'You would have to be very clever to understand that,' he said, 'because you would be understanding what is quite incomprehensible. So that's that.'

Presently she was back on the subject of his plays again.

'And two more, you told me, to be put on in the autumn,' she said. 'I want to hear every word about both of them. Begin right at the beginning, Hughie, and take them one at a time. What's the first one called?'

He got up.

'I shan't tell you a word,' he said. 'You would get it all muddled up with what I'm to read you after dinner. Now I think I shall go out for a stroll after all this strain.'

She wanted more of him and his doings, but much more eagerly did she want him to do as he liked. . . .

'Yes, dear, do,' she said. 'It will freshen you up.'

He knew she was longing to be asked to join him, but he made no such suggestion. They would dine and spend the evening together, and just now he wanted to get out of the range of her insistent questions. There were points, too, about one of these plays which he had promised to consider, and this could only be done in solitude. Tomorrow, perhaps, when she had heard and finished with 'Bread of Deceit', he would read her the second play. Yet he doubted about that; it was constructed on a triangular situation which she would not like. He could imagine her trying to throw herself into it, and, all the time, disliking it, shrinking from it. . . .

Then, some perception of her love, the royal gold of it, smote and upbraided him with a sense of his own bankruptcy. For very shame he must find one small bit of coinage. . . .

'But I'm not going out for a walk alone, am I?' he said. 'You're coming.'

The light leaped to her eyes again.

'Shouldn't I love to,' she said. 'But I thought perhaps that you wanted——'

'I want you,' he said. 'What else did I come here for?'

Three days passed, and there seemed to Hugh a great many hours in each of them, sunshiny, pleasant hours, with occasional clouds of tedium and irritation. He had many small polishings and filings to make to the play which would go into rehearsal as soon as he returned to London, and he lingered over these to the curtailment of those long sittings in the garden and slow strollings with his mother. He had read the play to her, and, as he had anticipated, she found the theme disagreeable. She did not seem able to consider it dramatically. She had to consider it morally, and her interest in the intricacy and solution of the situation was obscured for her by her dislike of it.

'Yes, dear, I see the power of it,' she said, 'and that by-play in the second act is quite delicious, sparkling with humour——'

'But?' he suggested.

'Oh, Hughie, it's so horrid,' she said ruefully. 'A man with any sense of right would have gone back to his wife. He would have cut

free of the whole thing when he saw where he was drifting. Couldn't
he do that? Couldn't he suddenly see the danger of it all? That
would make a splendid end.'

'But the whole play would vanish,' said Hugh. 'There wouldn't be
any play.'

'Oh, but there would! You might write a glorious third act making
us feel how he struggled against all his lower nature and conquered
it.'

'But he would have to be quite a different sort of man,' said Hugh.
'The whole problem hangs on his weakness.'

She looked puzzled.

'Yes, yes, I see that,' she said. 'But couldn't some other influence
come into his life?'

He saw the hopelessness of it.

'Yes, that would be all right, for another play,' he said. 'But it
wouldn't be this play. Now I shall go and wrestle with it alone.'

Certainly there was no use in discussions like this; it was a pure
waste of time to take part in them. She did not realize, so he told
himself, what a play meant; you could not have sudden ennobling
influences coming into weak and vicious lives and reforming them.
Tracts were the proper vehicles for such stained-glass emotions.
Morally, of course, it would be much better for everybody in a drama
to be noble at the beginning, or at any rate before the end, but
people did not go to the theatre to be edified: they went to be
amused and interested. And good people were not dramatically
amusing; there was nothing to be made of them. Domestic affection
was not material. . . .

A week later the two were installed in a furnished flat, and the
inevitable little frictions between two people of widely different ages
and tastes began their grinding; and the fact that the two were
mother and son contributed to, rather than diminished them. Mrs
Ranworth had indeed the inexhaustible cruse of her love to oil the
rubbing surfaces, and it was with the untiring joy of self-sacrifice that
she adapted herself to his utmost convenience. But with him it was
different; the very tenderness of her, the very eagerness with which
she tried to understand his needs and to enter into his life, jarred
and grated on him. All these years of toil, so long fruitless, all the
grimness of his determination to succeed, seemed to have given him
some sort of crustacean integument, hard and impregnable. Success,

the bleak fact of success, had been his sole preoccupation, and the pursuit of that had taken the taste out of all else. And withal even in the first flush of it there had come the deadly sense of its leanness. He had wanted nothing except that, telling himself that when he got that he could relax and care for lovely things again. But now that he had it, he was powerless in the grip of it. Things which his soul told him were lovely had become to him only material to be used in fresh successes. It was his business, and nothing else seemed to be his business except to dissect and to try to detect and to analyse. He could not drop the scalpel, and there was his mother for ever trying to filch it away from his fingers, to make him melt and yield, and open his soul to her.

It was not only in the affairs of his affections that he had become encrusted with this shell. In his struggle to obtain such a footing in the theatrical world as he now so surely occupied, he had devoted himself with unwavering energy to the study and production of what 'paid'. That had been his exclusive aim, and now with relentless justice that aim had stifled all others: it had crept over his whole mind like some monstrous growth of mildew. He could not now imagine himself spending his energies over any dramatic achievement which did not contain the elements, as he conceived them, of popular success.

And thus now he found himself deciphering the items on the bill which told him what had been the cost of success. He had paid it, anyhow, for here in the knowledge of the change which the hard years had wrought in him was its attested receipt. He could not any longer realize what it would feel like to possess the perceptions, the aspirations, the affections which he lacked. Their absence left no void: the soft mildew had filled it up.

His thoughts went back to his mother, and he was disposed to applaud himself for his aptness and ability with regard to her. A hundred times a day, her eagerness to fit herself into his life, her atmosphere generally, all she did and said, her very presence in the room irritated and wearied him. But never had he given her an impatient word, never had he failed in courtesy or consideration, and he fancied that as far as she was concerned, this joint ménage was an overwhelming success. To his blindness not a cloud seemed to sully her content: it was impossible she could guess how her very affection got on his nerves, so richly did he abound in the tokens of it. He hired a motor for her use, he took her to theatres, he made her little presents, he lavished on her the husks that contained no grain.

The new play was to be produced first in the provinces, and he was going up to Manchester for the first performance. The rehearsals had been heavy work, for while they confirmed him in his belief that he had another popular success, they showed him the deadly cheapness of it all. Yet he looked forward to his expedition, chiefly, he was aware, because he would be away from her.

'Perhaps I shall not be able to get back tomorrow, mother,' he said to her, 'for the first performance may show that we must make some cuts or additions or alterations, and I must attend to them. You can never tell from rehearsals exactly how the thing will play: you've got to feel the public's pulse. I dare say I shall have to be there for several performances.'

'Yes, dear, I quite understand that,' she said. 'Awfully interesting it must be to listen to the effect on your audience. Oh, I hope it will be a tremendous success! You won't forget to telegraph to me.'

She clung to him as he kissed her.

'Hughie, you've done everything possible,' she said. 'I've never been so deliciously looked after. You've . . . you've been delightful to me. All success, my darling!'

He duly sent her a highly satisfactory telegram. The first performance went so well that there was really no revision necessary, but he stayed for a couple of nights more before returning. His mother apparently was out when he got back, and with some sense of relief that he would not instantly have to embark on a detailed account of all that had happened, he went to his sitting-room with a handful of letters that were awaiting his arrival. On the top of these was a note in her handwriting, and guessing that it would only contain a line of welcome in case he arrived when she was out, he left it unopened till he had read such communications as promised more interesting matter. Among them was a weekly account of takings from the theatre where 'Bread of Deceit' was being played to full houses though the month was August. Other theatres were half-empty in this slack time of the year, but there was no languishing here, and the advance booking, so the secretary told him, was all it should be. Another letter concerned the third play; another was from a house-agent containing an order to view a flat which he thought might prove suitable. . . .

Finally he opened his mother's note. It ran thus:

MY DEAREST BOY—I have gone back to Blakeney in your absence, and I feel sure I have acted for the best. I could never have explained to you, with your dear eyes looking at me, why it is far better that I should leave you, for I should simply have broken down with all the yearning that is in my heart, and all my love and my gratitude to you. My darling, you have done your very, very best, but you don't want me to be living with you, and I can't therefore bear to do so. You have done everything that kindness and generosity can do, and I should be a very selfish mother if I inflicted myself on you any more. Naturally, quite naturally, you want to be independent.

If you think it over (probably even without that) you will see that I could not do otherwise. I knew very soon that my presence fidgeted you, and that was intolerable to me. It was unwise of me, I think, ever to have let you try the experiment, but you mustn't blame me for that; I simply could not resist it.

And now, my darling, let us just cut these last weeks out of our memory. I shall be back at Blakeney this evening, and oh! what a welcome you will get whenever you feel inclined to come down, from your ever most loving MOTHER.

He read it through twice, tore it up and sat for a while thinking.

'There's a situation in that,' he said to himself. 'Something might be done with it.'

ODD
STORIES

THE SUPERANNUATION
DEPARTMENT AD 1945

MANY people, of whom I am one, have from time to time, if they are given to dreaming at all while they are asleep, dreams which somehow seem to be of an entirely different texture from the ordinary nightly imaginings with their blurred outline, the incon-sequence of the events that take place therein, and the utter unreality of it all to the waking mind. Every now and then a dream of different stuff is woven in the sleeper's brain: that part of it—the subliminal self, or whatever it may be—which never wholly slumbers, is vividly astir, sends its message through the sleeping brain like bubbles rising in still, placid water, rising equally and sanely to the surface in undisfigured rotundity. Such dreams seem, after one has awoke, to be still actual, and though they are not exactly of the same texture as past realities, they are exactly of the same texture as the conjectured and anticipated future. They do not seem 'to have been', but 'to be about to be'. And when such dreams visit the pillow of the present writer, he puts them down when he wakes, and gets a witness to subscribe his name thereto. The witness, of course, cannot vouch for the vision, but he vouches for the date. Thus, if any of these dreams (the record of them reposes in a red-leather despatch-box) comes true, I shall send such, neatly dated and witnessed, to the Society for Psychical Research, as an authenticated instance of Dream Premoni-tion.

At present they are all still unsent. But of them all there seems to be none so vivid, so likely (remotely, for the Society will have to wait a long time) to come true, as one which visited me a fortnight ago. It still haunts me with a sense of reality, in comparison with which the ordinary events of today seem dim and unsubstantial.

The evening before this vision occurred I had been dining with sober quietude at a small bachelor party in St James's Street, and walked home afterwards, for the night was caressingly warm and unusually fine, with a friend and contemporary. During dinner we had talked chiefly about the delights of the High Alps as a winter resort. After that we had played bridge in silence, and walking home, we had talked about Switzerland again. I can find in the memory of our conversation and in the events of the evening nothing which could have suggested in the remotest degree (except that I was among old friends) any part of the dream. I parted from my friend at the corner of Albemarle Street, where he lived, went on alone, went straight to bed, and immediately slept. Then I dreamed as follows:

I was dining at a small bachelor party in St James's Street, and all those present—it was a party of eight—were well known to me. But our host, a very old friend—the same man with whom I had actually been dining the evening before—had been somewhat silent and preoccupied during dinner, and as we stood about afterwards, before settling down to easy-chairs or cards, I asked him if anything were wrong. He laughed, still rather uneasily, at this.

'No, not that I *know* of,' he said with rather marked emphasis. Then he paused a moment. 'I don't see why I shouldn't tell you,' he said. 'It is only that the Superannuation forms for the year have been sent out today. I was down at the Home Office this afternoon— Esdaile told me. Well, there are eight of us here, all old friends, and, you know, we are all of us over sixty-five.'

Now, though that fact had not suggested itself before, it was quite certainly true, and it was quite certainly as familiar as a truism. We had all of us got old, but the process had been natural and gradual. From which, incidentally, I gather that age comes kindly and quietly. Certainly the truth of his remark was apparent; there were only bald heads and grey heads present, and from where I stood I could see the reflection of my own in the glass over the mantelpiece, the shiny forehead reaching up to the top of the cranium, gold-rimmed spectacles, and grey eyebrows. Yet this—so vividly natural was the dream—was no sort of shock; that was the 'me' to which I was perfectly accustomed.

But his words, I am bound to say, were of the nature of a shock, for though for the last twelve years I had known that the annual sending out of the Superannuation forms might very intimately affect my contemporaries and me, I do not think I had ever realized it before. The circle of my friends was, I consider, large, though it was all present at that moment in this room, yet a man of seventy-seven who has still seven friends is, I hold, very enviable. But I could ill spare any of them; also, I could ill spare myself. All this passed in a flash, and since the mention of the subject was rather like a deliberate pointing to the Death's Head at a feast, I proceeded to turn my back on it. The Death's Head was there, we all knew that, for when eight very elderly gentlemen meet together at the time of the sending out of the Superannuation forms, there is always present the knowledge that they may not all ever meet again. But that, after all, is invariably the case. Anyhow, so I determined, I was going to enjoy the evening as usual. If this was to be the last time that this particular party, old and stupid and bald as we might be, were going to enjoy, as we had done for the last fifty years, each other's society, so much the more reason for making the most of it. If, on the other hand, we were going to enjoy it again, there was no reason at all for disturbance. So—I was a sprightly old man, I am afraid—I laughed.

'Come, let's play some old-fashioned game,' I said to our host—'bridge, for instance; let's play bridge and pretend we are all thirty and forty again. But we must play it seriously, just as we used to, in the spirit of forty years ago, when we all used to get so excited about it. By Gad! I nearly quarrelled with you over it and cut short a friendship that has lasted forty years longer.'

Now the knowledge that the Superannuation forms had been sent out had penetrated over the room, and out of the eight present there were certainly three rather grave faces. But the notion of playing bridge, a game that had been obsolete some twenty years, and of thus artificially putting the clock back, met with marked success, and in a very few minutes two tables had been put out. There was a certain amount of recollective disagreement as to the methods of scoring, but our host happily found, on a shelf of rare old books, a soiled and somewhat battered copy of the Rules of 1905 (first edition), in which year, apparently, certain small alterations came into force. With the shabby volume as referee, from which there was to be no appeal, we started on this queer old game, which always seemed to me to have certain good points about it, though now it was hard to get a rubber together, unless, as in the present instance, a party of elderly old friends were dining together. For myself, I cut the lowest card but one, and so—the copy of the Rules of 1905 upheld this—I was dummy.

Being dummy, and the first hand being a somewhat uninteresting declaration of clubs, it was not strange that I went back in my mind to the news I had just heard. And to make this dream vivid to the reader in at all the same degree as it was to me, I must enter into a short exposition as to my own feelings and habit of mind, as they were mine in the dream, in order that what follows may be intelligible. It is as vivid to me now—that outlook on life, and knowledge of the modes under which life was passed—as is my present outlook and the present modes of life to me now, as I sit here in the dim noon of a London day and write about the other from mere recollection of a dream. The year then was 1945, because I knew I was seventy-seven years old, and being that age I looked on life in a way that I can remember now with clear-cut vividness, though it was quite foreign to me. I looked, in fact, backwards, and my thoughts were as much and as pleasantly occupied with the past as they are now with the future. But this mention of the Superannuation forms distracted my mind both from the bridge that was being played, and from its habitual grazing-ground in the past, and made it wonder what risk any of those present (and, in particular, myself) ran of receiving one. The whole system of the Superannuation scheme was, of course, perfectly familiar to me, and though in this year 1905 it seems to me rather brutal, it did not seem so in the least in my dream. Familiarity with it may partly account for that, but what more accounts for it, to my mind, is that in the year 1945 one looked on the mere fact of life (the tenses are difficult) in a manner altogether different from that in which one looks on it in 1905. In 1945 the life of the individual mattered far less than it does now, or—which, perhaps, is the same thing—the life and well-being of the nation mattered far more. This, I think, is one of the probable points about the dream, and to my waking mind it was Japan and her heroic, unquestioning sacrifices in 1904 and 1905 during the Russian war, which began to wake the Western nations up to the undoubted fact that to progress as a nation the individual must sacrifice himself by his thousands (or be sacrificed) without question or demur.

Briefly, then, the Superannuation scheme was this. Anyone over the age of sixty-five was liable to receive each year from the Home Office a printed paper, which, like the income-tax return, he had to fill up to the best of his power and belief. Everybody over that age did not receive them, but a very large number were sent out each year. In this paper were some eight or ten questions, as far as I remember (I shall not forget them or the number of them again), and, to certain of these, witnesses—who were liable to have to swear to the truth of their testimony, and were subject to cross-examination—had to append their names. And if, in the opinion of the Board for Superannuation (attached to the Home Office), the answer to these questions was unsatisfactory, the returner of the form 'died' within a fortnight. This Board for Superannuation consisted of the most humane, wise, and kindly men, and any of those who were related to the filler-in of any particular paper, or who could, in the most remote manner possible, profit by his death, were debarred from adjudicating or voting in any such instance. I had several friends on the Board; indeed, I had once been asked whether, if a seat there were offered me, I would take it. This I had declined. The manner of death was infinitely various, and reflected great credit on the ingenuity of the contrivers. It was also perfectly painless, and, I believe, even pleasant. Such was the sum of my musings about the matter while the hand of clubs was being played.

Now all this seems somewhat cold-blooded and unwarrantable to us in 1905; but in 1945, owing chiefly, I think, to the utterly different value put then on mere life, it seemed perfectly reasonable. The population of the world had, of course, vastly increased, and there was no ground left for useless people to cumber. The law had been in force some twenty years, and the form drawn up with the most scrupulous care. Any valid cause why a man should continue to live was cause enough. What exactly the questions were I did not at the moment remember. Afterwards——

However, for the present the bridge went on, and it was late when this pleasant though elderly party broke up. The night was warm and fine, and I walked home with a 1945 edition of the friend mentioned above, with whom I had walked home in 1905. Old times, as usual, occupied our thoughts, and we recalled our fifty years of friendship with no little complacency.

'And half-a-dozen times, at least, every year,' said I, 'we must have walked home from that door together. Three hundred times, at least. Well, well!'

'And three hundred times, at least,' said he, 'I have asked you to walk a shade slower, just a shade slower. All these fifty years you have never mastered the fact that I am two years your senior. Well, I turn off here,' he added, as usual, at the corner of Albemarle Street. 'Good night, good night. See you at lunch at the club tomorrow?'

'Rain or fine,' said I (also as usual).

Now, to younger people this all sounds very dull; just two old men of near eighty who had often and often bored and irritated each other, toddling home, and settling to lunch at the club next day. But there seemed to me then in the dream (and, indeed, there seems to me now, when I am awake) a certain humanity, a certain achievement in the mere fact that these two old things had preserved their tolerance and liking for each other during so many years. I am glad to think that I was one of them, for they must have had rather kind hearts and a pleasant indulgence for each other's irritating qualities. In fact, I sincerely hope that this part of the dream may come true.

I let myself into my flat and went into my sitting-room to see if there were any letters. There was only one, in a long, pale-yellow envelope, unstamped, but with O.H.M.S. printed at the top. It looked like income-tax. It also looked like something else.

I opened it; a small white printed paper fell out and fluttered to the ground. There was also a long, yellow printed paper with many blank spaces in it. I read the small white paper first:

> Home Office, Whitehall.
> May 9, 1945.

Sir,

The Board of Superannuation beg to enclose the usual form, with the request that it may be filled in according to the instructions, and returned to them within the space of seven complete days. For every additional day beyond these you are liable to one year's imprisonment as a criminal of the second class.

Should your return be satisfactory, you will be informed of the fact within fourteen days of the receipt of your return.

> I beg to remain,
> Your obt. servant,
> A. M. AGUESON (Secretary.)

Then I read the other paper:

O.H.M.S.

———

Superannuation Department

———

The recipient is required to fill in answers to the following questions to the best of his ability and belief.

Witnesses are liable to be called upon to repeat their testimony on oath and subject to cross-examination. Suspected perjury on this point will subject them to criminal prosecution.

I. Are you useful?

(Useful is taken to mean *productive* in the widest sense of the word. The answer should therefore include (a) any works or objects of art which the returner is in the habit of producing, (b) all scientific or other research work on which he may be engaged, (c) any other pursuit in which he is now personally engaged which, in his opinion, adds to the pleasure, wealth, or happiness of the nation or of individuals.

Sub-section (d). Mere employment of labour or mere contribution to charities does not fall under the preceding heads, unless such is accompanied by active work, investigation, or enquiry on the part of the owner or donor. Witnesses to the answer must be: (a) art-critics of the specific art in question of recognized standing, (b) scientific men, (c) responsible manufacturers, and [sub-section (d)] commissioners of charity organization or similar and recognized schemes.)

Answer   *Witnesses.*

II. Are you beautiful?

(Beautiful must be taken to imply an object of positive beauty, the contemplation of which is calculated to afford artistic pleasure to the beholder, and stir the artistic into production.

Witnesses to this section must be professional artists, two at least in number, of the standing of A.R.A.)

Answer   *Witnesses.*

III. Are you morally better (though still, perhaps, bad) than you were a year ago?

(Honesty, temper, tact, good nature, patience, truthfulness, content, are all reckoned moral qualities.

Witnesses (not less than three in number) must be (a) clergymen of the

Church of England in priests' orders, or two bishops are considered the equivalent of three priests, (*b*) domestic servants.)

Answer   *Witnesses.*

IV. Are you contributing in other ways than by moral worth, personal beauty, etc., to the reasonable happiness of others? If so, how?

(The word 'happiness' to be taken in its broadest sense.

Witnesses to the answer should be not less than three in number, and consist of those who most habitually see the signatory—i.e. friends and domestic servants. The signatory is also recommended to note with the greatest possible accuracy (since this will be tested) the effect that the news that he has received the Superannuation form makes on such).

Answer   *Witnesses.*

V. Are you likely to become an object of beauty?

(Enclose two photographs, if an affirmative answer is returned, (*a*) of this year, (*b*) of any previous year. These photographs will be returned by the Home Office in any event. No witnesses required.)

Answer

VI. Are you happy? If so, give a brief sketch of your average day, stating from what your happiness is derived. No witnesses required.

Answer

VII. State broadly any additional reasons you may have for wishing to continue to live. No witnesses required.

Answer

(This form must be folded and sent entire within seven days. No stamp need be affixed.)

It was as I read through this that, for the first time, any sense of nightmare or horror awoke in me, and as question after question conveyed itself to my mind, this horror gained on me. I could not say I was beautiful; at least, I could not get an A.R.A. (still less two) to agree with me, except at very grave risk of their incurring the penalty of perjury. Or what three clergymen would say I was better than I was last year? But on purely personal grounds I wanted to live. No doubt that was unworthy; my room, no doubt, was more useful to the nation than my company. But I still wanted to live, and as I came—so I suppose—nearer to waking, I more and more wanted to live. Whatever the past had been, whatever was the present which was constructed on that, I wanted the future and its opportunities. My own live self, in fact, as my sleep became less deep, began to grow more dominant, while the aged 'me' of the dream began to fade, till, with a strangled cry, protesting against the wild injustice of being put out of the world, I awoke, with flying heart and perspiring head, to find my room bright with the newly risen dawn and all the promise of another day.

Now, never in all the archives of this leather box have I had a dream so distinct with the sense of sober reality as this; and as the days passed on, that reality grew no less, till now, when a dozen days have passed, I can recall, as vividly as I can recall anything that ever happened to me in waking hours, the sense of being old, the sense, too—which is utterly alien to me—of looking backwards instead of forwards. For up to a certain time of life one is like a traveller who is seated facing the engine, and ever looks just ahead of what is immediately opposite him. But that time past, for fear of draughts or what not, we gather up a railway rug, seat ourselves with our backs to the direction of progress, and see only that which has passed us.

Again, though the perturbation of waking woke a sense of rebellion in the dreamer's mind as to the justice and expediency of the Superannuation scheme, my belief in it now is fast and firmly rooted. For—such is the wisdom of the questions—no one, except the most useless drone, stands within the danger of the State-inflicted death. Usefulness, beauty, cause of happiness in others, improvement in oneself, even mere personal happiness, are all taken to be signs—or so I read the paper—that the signatory of the form is still paying his way, so to speak, in the world; that his presence there, being a source of encouragement and pleasure to others, is still desirable; that he is still in some sense a growing being, not a mere blind block on the highway of life over which others may trip and hurt themselves, and which is far better removed. In every line of this dream-document there is statecraft, and in none more clearly than in the clause that distinguishes between mere employment of labour, mere charitable munificence, and real usefulness. For such employment of labour and such munificence is but a mechanical function, and could be as well, and probably better, done by others than by one who in no other way contributes to the national welfare. That clause, in fact, seems to me really Japanese in point of insight.

Further, how wise is the question: 'Are you happy? If so, why?' For here the State recognizes that innocent and instinctive happiness is in itself a gain, a dividend-earning proposition. For happiness is as infectious as misery (which is saying a great deal), and a happy man cannot help contributing to the welfare of the world. It is a fact not yet properly recognized, and I rejoice to know that in 1945 it will be.

Again, in those Utopian days, it will be recognized that beauty is a contributor to the welfare of nations. It must be allowed that now, while London is London and, more especially, New York is New York, a great gulf is fixed between now and then, as regards our Western civilization, where county councils and other bodies of high intelligence are steadily employed in substituting the ugly for the beautiful, wherever such substitution can be made without undue expense or sacrifice of efficiency. But in 1945, so I have reason now to hope, even though beauty be of so senile a quality as may be exhibited in gentlemen of sixty-five and over, it will be recognized as an asset in a nation's solvency and a reason why the possessor of it should be permitted to live. And from where but from the East may this dawn be expected to enlighten the skies? Here, again, Japan springs to the fore—Japan, who in the midst of the most sanguinary and expensive war that the world has ever seen, celebrates with her accustomed courtesy and merriment the festivals of Chrysanthemum and the Flowering of the Cherry.

Again, how wise and 'insighted' to make mere domestics competent witnesses as to a man's habit of diffusing happiness, a thing so vastly important; while for the mere support of his claim to beauty, A.R.A.s are required to give their signature! For this seems to be at last a practical recognition of the truism that charity begins at home. Deeds of trivial domestic kindness, and the habit of them, are recognized at their real value in this dream-document. Mark, too, the severity of the punishment for perjury.

On first consideration the penalty for delay in sending in returns seemed to me disproportionate to the offence, but on subsequent reflection I think it is right. For any man who dallies with death for the mere sake of living another day is no longer fit to live, being an essential coward. And if we want to get rid of the superfluous population, let us by all means begin by segregating and putting in confinement all essential cowards. For really there is no use for them. Cowardice stains the whole character: it eats like corrosive acid into whatever apology for other virtues there may happen to be, and renders them futile.

Finally, how sound a principle underlies the whole scheme! Such a paper might indeed be sent with advantage, not merely to poor old folk of over sixty-five, but to all adults, since its challenge is 'Justify your existence.' If any man cannot justify his own existence, it is almost certain that nobody else can do it for him. He came into the world through no volition of his own: surely he may be enabled to leave it in the same manner, if his presence there is unjustified on so broad a field of enquiry as is covered by this Superannuation form. Above all, if he is not happy, he will not be sorry to go, while if he is, any reasonable grounds will be accepted by the Board—or so I read it—as a sufficient reason for his being allowed to live. But—this, too, is wise—the grounds of his happiness must be reasonable. I cannot imagine the Board accepting a burglar because he took pleasure in stealing.

So there in the leather box this dream reposes. It would give me great pleasure—if it were in my power to do so—to dream on the same subject again, in order to clear up, for my own satisfaction, several points which are still vague to me. I want to know, for instance, whether one affirmative answer, if completely satisfactory, entitles the signatory to a fresh lease of life.

Ah, yes, it must be so. However hopeless in other respects, a man of over sixty-five who can thrill with joy (and satisfy the Board on the point) when, on an early day of spring, he sees the pale crocuses peer above the grass, and feels the spring in his bones, is surely worthy to live, on the mere consciousness of his own happiness, whether he be twenty years old, or seventy, or ninety—in fact, the older he is, the less he can be permitted to die, if he can possibly be kept alive. For on such a day, though it is easy for the blackbirds to have their will, it takes a poet to have his.

THE SATYR'S SANDALS

WHEN Pan signed his contract for himself and his company to reside in London, he had made it terminable, without notice, by either of the high contracting parties, of whom he was one and the Genius of Piccadilly Circus the other, for he knew perfectly well that nobody could ever want him and his delightful Satyrs and Dryads to go away, whereas he and they might conceivably get tired of town. He had also bargained for one day off during the week, and as was natural, he took it on Sunday, when the London folk grew rather yawnsome in the evening from being bored all day, and did not so eagerly respond to the stimulus of his chaperoning.

The Dryads and Satyrs who were, so to speak, the ladies and gentlemen of his most democratic court, followed his example, and in the summer you could often catch a glimpse of a white shoulder or a pointed ear among the crowds at Richmond, or further afield in Boulter's Lock above Maidenhead. But as they all had to be at work again early on Monday they usually came back during the night by some late-returning motor, or lying among the flowers and vegetables that came up by van to Covent Garden. Being so light they hardly bruised the ripest strawberry, and the horses made the journey with far greater speed and willingness if they were conscious of their delicious presences. Then they slept casually in the fountains in Trafalgar Square or along the edge of the Serpentine if it was warm, smelling of buttercups and hawthorn and dog-roses, till it was day. It often happened that these simple fragrances of childhood went to their heads after a fine day in the country and they forgot the time, so that on Monday morning when Pan read the roll-call from the fountain in the Circus, with the Genius standing by to check the names, there were some absentees who came in very much out of breath with lamentable excuses. Sometimes one or two of them would play truant more seriously, and once at Cookham I distinctly saw a Satyr at eight o'clock on Monday morning in hot pursuit of a Thames water-nymph. I think he caught her because I heard shrieks of Sunday-night laughter, and the sound of splashing. . . . In the winter many of them did not leave London at all, but dozed behind statues and cabmen's shelters, or, if it was really cold, slipped down into the silent, stuffy recesses of the Tube, or lay round the expiring embers of grillroom fires.

On this particular night towards the end of February Pan came clattering up Lower Regent Street. It was very mild and moonlit, and as he went through St James's Park he had seen several of his court on the rock where the pelicans roost, giving them nice dreams about fish. His hoofs made such an echo against the blank walls of the houses, that a Dryad and a Satyr who were sleeping just outside the Picadilly Tube Station woke up thinking that they heard the rattle of the early traffic. They had only gone to sleep by accident, and when they saw that the noise that had awakened them was Pan, they were quite pleased to have their slumbers interrupted. So, when he came bucking round the corner, the Dryad just smoothed her hair by way of assuming levée-dress, and the Satyr fastened the string of a sandal which had somehow come undone. He could not make a more elaborate toilet, because he had nothing else to make it with.

'Don't bother to salute,' said Pan. 'What have you two been up to?'

The Dryad yawned in a feminine manner (she had caught that odious trick in London) so as to make the two gentlemen think she was bored, in which case politeness and chivalry demanded that they should exert themselves to amuse her: her yawn was merely a signal for their activities.

'Oh, it's been devastating,' she said. 'I hate Sunday.'

'You're very rude,' said the Satyr, 'considering you've spent it entirely with me.'

'Darling, I didn't say I hated you,' she said. 'What I hate is that the humans should be so dreary once a week, and really they are not much better now on the other days. I rather wish you signed on for Paris and not London, Pan.'

'With the franc at forty-five?' said Pan. 'Don't talk to me!'

'But with all respect, dear, I do talk to you,' she said. 'I quite see why you signed on for a town, because just now all the gaiety in the world goes on in town. It didn't used to——'

'Didn't used!' quoted Pan indignantly. 'What a Cockney you've become! You ought to be happy here.'

She gave a little indulgent sigh.

'Used not to,' she corrected herself, 'though what you call grammar, I call pedantry. I understand why you signed on for a town. But why London?'

'You have my permission to go to Manchester, tomorrow, if you like,' said he.

He clasped his arms round his shaggy knees.

'I chose London,' he said, 'because I thought that when the war was over London would be the very centre and climax of the joy of the world. London won the war——'

'Say, stranger'—began the Satyr, through his impertinent nose.

'Now no more about New York,' said Pan. 'Just be thankful you are not there. People are sadder and busier there than anywhere.'

He loosed his hands and scratched his ear with his hoof.

'I confess that London isn't what I thought it was going to be,' he said. 'Shall we terminate the contract and go to Paris?'

The Dryad leaned her pretty head back against the Satyr's shoulder. He had kicked off one of his sandals again, because he thought more vividly with bare feet. The other, which he had just tied, would not come off.

'What do you say?' she asked. 'I don't really know that Paris would be better than London. And then there's the exchange.'

'Oh, bother the exchange!' said the Satyr. 'It isn't what people spend that matters; it's the spirit in which they spend it. Last week I went down with a bank clerk and his girl to Brighton. All they spent was a few shillings on shrimps, and a third-class return ticket. But they were gay; I enjoyed it much more than I have enjoyed many of these restaurant dinners in town, followed by a dance at Cantogalli's and supper afterwards.'

Pan slipped over him as he lay there, and insinuated himself between him and the Dryad, putting an arm round the neck of each. He hated with all the force of his nature that his young folk should be bored and blasé.

'Put your heads back and make yourselves comfortable,' he said, 'and tell me what's the matter with London. It's ailing somehow, and so are we all. Here's one of my Dryads yawning, and one of my Satyrs saying that Cantogalli's is a bore. Now what is it?'

The Dryad nestled up close to Pan.

'I feel old,' she said; 'I even feel middle-aged, which is worse. When you're feeling old, you know you can't be young any more, and so don't try. But when you're feeling middle-aged, you keep on trying and are always disappointed. And I want to feel young again: I want to feel as I felt when I used to go bright-eyed through Athens all night long. Was it the air that made us young there?'

'No, dear, it was you who made the air young,' said Pan.

'Then why can't I make this soupy London air young?' she asked. 'I can't. To think that tomorrow is Monday, and that once I used to look forward to Monday so enormously, because all the week was before one. Now I don't care whether Sunday goes on for ever. Oh, what shall we do? Shall we dance, Pan? I used to love dancing with you, you dreadful old flirt!'

'You don't want to dance,' said Pan.

'I know I don't. It's awful for a Dryad not to want to dance. It's worse than not dancing. I'm getting like human girls: I'm slack and tired: I don't care. . . . That's what's the matter with me.'

'Same here,' said the Satyr sleepily 'Everybody is bored. They dance because there's nothing else to do. They yawn because there's nothing else to say. They drink because they don't want to eat. They eat because they don't want to drink. What's the matter with them? And why should we feel like that too?'

Pan drew his Satyr and Dryad closer to him.

'*Götterdämmerung!*' he said.

'Don't talk German,' murmured the Dryad.

'Very well. It's the "Dusk of the Gods."' said he. 'For all these centuries we have thought that we pulled the strings, and that the human race danced to our pullings. But in reality it has never been so. It was they who pulled the strings, and we danced. We didn't make them: they made us. Once upon a time the shepherd boys played on the slopes of Penteticus, and the girls who tended the goats had soft, dark eyes, and joy was born. And because they did not know that they themselves were the fathers and mothers of joy, the girls invented this young slim god of the thickets, and the boys invented this nymph of the woodland. And then they must needs have a lord and master of you two, and they invented me. I wasn't there till they thought of me. Then Nature gave us a life of our own, so that men cannot destroy us whom they made. But now we have grown dim to them, because they are tired and have forgotten joy. So let us be off, children. When they find we are gone, they will begin to long for us again, and they will run out of their town to seek us.'

He raised himself, lifting the two beautiful drowsy heads that lay on his shoulders.

'Come back to the great Mother,' he said. 'In the country the wild things, birds and plants, and the little furry beasts are all a-tremble with the coming of the spring. The buds of the sallows are bursting, and the streams are full, and the hazels droop with catkins. Go on sleeping, tired ones, and have dreams of old days, and you will wake to find them coming true.'

He folded the languid bodies across him, and, as an aeroplane gathers speed before it lifts, he galloped down Lower Regent Street, and floated off into the air. He rustled through the tree-tops in St James's Park, cleared Queen Anne's Mansions with a foot to spare, and headed south by east.

Not five minutes after he had gone, I, unable to find a taxi, and hoping against hope that the Tube was still running, came to the threshold which they had just left. To my amazement I saw a Greek sandal on the pavement, which I picked up and carried home.

Next day I went to Rye, and arriving early in the afternoon went for a walk in the marshes. Just where the hill of Winchelsea rises from the level land, I passed by a copse all riotous with spring. The sallows were studded with their moleskin buds, the catkins were pendulous in the hazels, and the primroses were already in flower. More remarkable even, since February had still a week to run, were the carolling larks that rose from the tussocks, and most remarkable of all was a thing that caught my eye on the edge of a reed-bed. It was a Greek sandal. . . .

THE DISAPPEARANCE OF
JACOB CONIFER

MEMORIES nowadays are short, and the rapid succession of journalistic excitements quickly expunges from the minds of readers the record of previous thrills, but I expect that, though some time has elapsed since the disappearance of Mr Jacob Conifer, most people still remember the prodigious interest it roused throughout the length and breadth of England, and the United States, of which country he was a citizen. Friendships were dissolved between disputants as to the key of the mystery, elder sons were cut off with shillings, lasting animosities kindled, old ladies left their fortunes to their canaries instead of their companions. Armies of detectives were employed in following up clues which invariably led to nothing at all, and the most ingenious gentlemen from Scotland Yard were baffled, for whatever theory they constructed, some fatal objection to it cropped up, and never did a particle of success attend their patient investigations. The brains of the whole country were roused: an ex-Lord Chief Justice, steeped in criminology, filled two columns of the most widely circulated paper in England with arguments to show that Jacob Conifer had been murdered, a professional billard-marker and a Bishop convinced many thoughtful persons that he had committed suicide, and the Secretary for Foreign Affairs urged us all to believe that he was still alive. Blackmail, robbery, loss of memory, all had their acute supporters, but out of so many theories, evolved by distinguished intellects, not one really covered the facts, and no subsequent discovery has justified any of them. The riddle still presents (or did, till yesterday evening) a Sphinx-like face to all who attempted its solution.

But last night, as I dined alone with an ingenious friend, who, like me, knew something about Conifer, we went through very carefully our recollections of the vanished personage, with all the circumstances of his vanishing, and between us we hit on a perfectly new theory, which, fantastic and insanely incredible as it sounds, covers, as far as we could see, absolutely all the known facts. What is more, it accounts, though still perhaps incredibly, for some very odd things concerning the personality and psychology of Mr Conifer, about which there is no doubt. I will first, therefore, narrate a few facts about Mr Conifer which are not disputed, recount the circumstances of his vanishing, and then lay before the reader, in a mixture of diffidence and pride, this astounding theory of ours, just as it emerged in our conversation.

Mr Jacob Conifer made his meteoric appearance in England three years ago. He appeared to be immensely wealthy, and was one of that amiable company of Translatlantic cousins whose mission in life appears to be to make the English folk of London acquainted with each other and with them. These hospitable people are usually women, but Jacob Conifer, unique in his life and death (if he is dead), was a man. After a very short chrysalis period at one of the central London hotels, which he occupied in leaving cards and letters of introduction, he bought a house in Grosvenor Square, a moor in Scotland, a mansion at Newmarket, a yacht and some golf-clubs. He made his mistakes, as shall be mentioned later, but nothing caused him to swerve from his purpose, or indeed seriously retarded his progress. Forcible feeding was his method with London: if he wanted anyone to come to his house (and he wanted everyone), he went on asking them till they came. Many resisted for a while, but sooner or later he won; it was a question of moral brute-force.

The end was quite sudden. He had asked a large party down to Newmarket for the July meeting, and left London on the afternoon of the previous day, having ordered his motor to meet him at Cambridge. His luggage and servants had all gone on, and he travelled with a despatch-case, which contained a Peerage, without which he seldom moved. Conifer had been out to lunch that day, and went straight from the house to the station, arriving only a couple of minutes before the train was due to start. On going to the booking-office he found he had only just enough money to pay for a third-class ticket, and his frenzied assertions that he was Jacob Conifer left the clerk at the booking-office quite cold. He had to go third-class or not go at all. The train was absolutely packed and he was thrust into a corridor choked with the proletariat, and the door with difficulty shut behind him.

At the first stopping place there was some commotion, and a man, apparently in a state of collapse, was carried out of the corridor: his luggage consisted of a despatch-case. The two porters who laid him down on a bench in the station, deposed, at subsequent interviews, that he had a strangely shrunken and withered appearance. He was unconscious and evidently extremely ill, and medical aid being summoned, he was at once conveyed in an ambulance to a ward in the Infirmary, and put to bed after being undressed and clothed in a set of parochial pyjamas. His identity could be established later; the first thing to do was to save his life. Presently he recovered consciousness, and in a very feeble voice, he said to the nurse, 'Where am I!' She told him reassuringly that he was quite safe in the Infirmary, which did not have the comforting effect she anticipated, for the unknown patient sitting up in bed gave a wild and awful look round the drab but sanitary room, and on his neighbours, and lay down again with a convulsive movement, covering his face with the sheet. In accordance with her orders, which were that she should summon the doctor if he came to himself, the nurse hurried off to find him. There was some delay, for he was engaged with a ticklish case of delirium tremens, in which he urgently required assistance, and it was not for ten minutes that she returned with him. Conifer's bed had looked strangely empty, and they found in it the suit of parochial pyjamas, but of him no trace whatever. A thorough search of the ward and the adjoining rooms and passages was made, and the other occupants were questioned. But no movement, they said, had come from the bed of the missing man, and the search revealed no trace of him. His clothes, too, were on a chair, all complete, and it was evident that he had left the ward without a stitch or a shoe on him. But the examination of his clothes brought to light two calling-cards: one in his breast-pocket was that of the Duchess of Middlesex, the other that of Jacob Conifer. They were sure he was not a Duchess, and the other card established his identity.

But since that fatal afternoon, nothing has ever been heard of Mr

Conifer. Apparently he had no trouble on his mind, indeed, he was much looking forward to his party at Newmarket, and his account at his bank showed a positively indecent balance. Even if he was dead, there could be no inquest, because there was no corpse, and, if he was alive, the conundrum of what had happened to him, baffled, as I have said, the choicest brains of our lay and clerical fraternity.

My friend and I talked all this over, and found our recollections of the case exactly to tally. He, being one of our most eminent diplomatists, does not wish his name to be mentioned in connection with the conversation that followed, for he says that the Foreign Office would distrust his sobriety of judgement, if it was known that he contributed to our fantastic conclusion. For the purpose of this report, therefore, I call him 'Jim'. It is, however, an essential part of our solution that the disgusted reader should be at once informed that his sister, Madge, married a Duke, and is therefore a Duchess. It was her calling-card, in fact, carried in poor Mr Conifer's breast-pocket, that had been discovered there. It was quite genuine: Mr Conifer had not had it privately printed.

So Jim and I ran over the facts of the disappearance and then, idly enough at first, began to survey our previous knowledge of Mr Conifer.

'He had a letter of introduction to me,' I said, 'when he first arrived, and I asked him to lunch: just he and I. A small queer little fellow like a shrimp——'

'What do you mean?' said Jim. 'He had an introduction to Madge, and she asked him to one of her football scrimmages on the staircase which, she thinks, advance the cause of Conservatism. I was standing near her, and remember seeing him come up the stairs, and wondered who that big man was. He was quite large: certainly over six feet. I heard his name, so it must have been he, but I wasn't introduced to him.'

'He had grown since I saw him then,' I remarked.

I had hardly said it, when I began to distrust my own impression of him, for the next time I came across him was at a bazaar in aid of some meritorious charity, which was opened by a Royal Princess. Conifer was hard at work by then; he had made a munificent contribution to its funds, and I suddenly visualized him, bowing low over the gracious hand and smile extended to him, and then straightening himself up. He did appear to be a big man, and I felt I ought to correct my first impression. But for the moment I was saved.

'Yet I'm puzzled when you say he was small,' said Jim, 'and I wonder if you are not right. I saw him once at Hyde Park Corner submerged in a crowd of people waiting to board a bus. Conifer had got wedged in among them, and he was wearing a tall silk hat, but I recollect quite clearly that I saw the top of his hat only just appearing. Perhaps he was small.'

'Well, if you confess, I must do the same,' I said. 'I saw him once bowing to a Princess, and he looked immense. I scorn to argue that it was a small Princess. Let's agree that both you and I thought he looked small on some occasions and big on others. It's queer, though.'

Jim was silent a moment, his brow knitted, his eye alert and yet lost in some silent speculation. But whatever it was he dismissed it, and laughed.

'How he besieged Madge,' he said. 'He asked her to dine every day for a fortnight. For ten days she refused, but on the eleventh she broke down. I was dining there, too, and he hadn't grasped our relationship, for when Madge arrived he solemnly introduced me to her. Madge was in her most devilish mood, for she gave me an icy bow, and turned to somebody else. She explained afterwards that she thought it would be *too* awkward for Mr Conifer if she told him that she was my sister.'

'But I was there,' I said. 'It was I to whom Madge turned and said, "What are we to do?" And it really is very strange: I thought he was big when he shook hands with Madge, and even as I was telling her that she ought to have laughed it off, and said how pleased she was to meet her brother, I looked at Conifer again. He was evidently most uncomfortable: he thought that she was a disreputable Duchess, or you a disreputable diplomat, and he looked small: he looked wizen and withered.'

Even as I spoke, some glimpse of what we were searching for gleamed on me. It was faint and evanescent, for before I had time to grasp it, it was gone. But I knew, somehow, that Jim and I were on the same track. . . .

He laughed again.

'Everyone behaved very badly to Conifer,' he said. 'If they wanted some lunch they dropped in at half-past one to his house in

Grosvenor Square, whether they had been asked or not. And if they didn't happen to want to lunch there on the day on which they had promised to go, they left it alone. Sometimes you found a table laid for twenty, with Conifer sitting solitary at one end, and sometimes you found a table laid for four, with twenty people trying to eat at it. And sometimes—good gracious, I'm beginning to see light——'

I sprang out of my chair: it was the same beam shining on us both.

'I know what you mean,' I cried. 'When Conifer was sitting miserably alone at a large table, he was little. If there were lots of the people he thrived among, he was big. I remember a lunch-party there, and I was the only man who wasn't a Marquis or better, as they say in jack-pots, and Conifer was immense. A huge man.'

'Now, don't let us be in a hurry,' said Jim, holding up his hand. 'I guess what is in your thoughts; you guess, I expect, what is in mine, Marquises, Duchesses, social success—'

'Made him large,' I interrupted.

'Stop there a minute,' he said. 'Let's see if it is consistent. He was big when he came up the stairs at Madge's party: he was immense, you said, at that bazaar, he was immense at that party you speak of——'

If my tongue had been cut out that moment, I should have gesticulated with the deaf and dumb alphabet.

'And he was small when he made a gaffe about introducing you to Madge,' I cried, 'and he was very small when you saw him in a crowd at Hyde Park Corner, and he was small when he came to lunch with me alone, and when nobody came to lunch with him.'

Jim looked at me a moment in silence, like Cortez on some peak in Darien.

'It almost frightens me,' he said. 'It's too stupendous. . . . The train was crammed the day he started to go to Newmarket; he was standing in the corridor all the way, wedged in with common people.'

'And when he was carried out of the train in a state of collapse,' I said, 'the porters noticed that he looked shrunken and withered. It all hangs together; social failure and the pressing of the proletariat upon him wilted him. I don't suppose he knew he got smaller; he only felt unhappy, and the physical change was the effect of the mind acting on the body. A very clever doctor told me the other day that the power of the mind over the body was limitless.'

Again Jim's eyes grew large with awe.

'The final scene!' he said. 'Just think of it! It is like some great Greek tragedy realizing itself in modern life. He recovered consciousness in a bare bleak room with whitewashed walls and carpetless floor, and to right and left of him were other beds with the bodies of pauper invalids lying in them. He was already much shrunken, you must remember, from that terrible hour in the train.'

I could not bear to be a listener merely.

'And then he asked that fatal question,' I said, 'and was told he was in the Infirmary of a country town quite unknown to him. It would be a shock to anybody. Think what it must have been to such a one as he! The humiliation overwhelmed him——'

'And he went out like a candle,' cried greedy Jim.

So here, for what it is worth, I present to the reader this novel theory concerning the disappearance of Jacob Conifer. It may strike him as entirely preposterous, but let him carefully run through the known facts, and consider whether it does not cover them all. Or let him study what the ex-Lord Chief Justice, the professional billiard-marker, and the Secretary of Foreign Affairs had to say about it, and then see if he does not prefer a theory, however startling, that agrees with the entire evidence, to any of their lame explanations. Let him remember, too, that Jacob Conifer was already much shrunken by his terrible hour in the train, and that by any other theory he would have had to glide naked and unperceived from his bed, traverse a corridor and a flight of steps, and emerge into the populous streets. But the invalid paupers were quite certain that he did not leave his bed, and he was never seen again outside it. . . . In fact, I passionately protest that Jim and I must have thought of the true explanation: I challenge the whole world to put forward another that so simply and comprehensively meets the case.

DODO
STORIES

THE RETURN OF DODO

'I DON'T care a pin what you say, my dear,' remarked the Duchess, who was in rather a hurry, to one of the vice-presidents, 'and all I know is that I shall ask her to open it on the second day.'

So the Duchess did ask her, by telegram, reply paid, to Paris; and Dodo replied, 'Charmed'.

As soon as her answer was made known to the committee—the Duchess had not consulted them in a body, as to the step she was going to take—the larger section, with a childlike faith in duchesses which did them credit, assumed that its success was assured, and congratulated each other on the certainty of finding themselves in possession of sums so enormous that they could guarantee to the public that no child should ever suffer from cruelty till at least the end of the century; for, as Lord Ledgers remarked wistfully, things always *used* to have a habit of humming when Dodo was there. A Royal Princess was going to open it on the first day, but the Duchess put all the arrangements for Her Royal Highness's reception into the hands of a deputy sub-committee; for while the Princess was, according to her, the investment of a small sum in consols, Dodo was an investment in a South African gold-mine—somewhat risky perhaps—and enormous and immediate returns might be expected.

Dodo had been out of England for two years, and it might have been supposed that London, or rather that momentous fraction of it called 'All London', would have entirely forgotten about her. But at the end of that time, her husband had been moved from Madrid, where he had been first secretary to S.M. the Austrian Emperor's embassy, to London, as S.M.'s First Secretary there; and as soon as this fact was made known, Dodo's name leapt to All London's mouth. One section said that it was impossible that she should come back; they would lay long odds that in a few days it would be announced that Prince Waldeneck had declined the post, owing to the ill-health of his wife, and the necessity for her of a Southern climate—here they winked at each other, and said that a climate remote from London was what *she* was in need of—London would not suit her at all. Another said he would come, but that she would not. Another that they would both come—it would be like Dodo to come—but that each individual door of All London would be closed against her. Another that Dodo had—well, they would not call it reformed—but given up all that made her amusing before, and now spent her time in making altar-cloths and knitted comforters. (They had it on the best authority, having made it all up themselves.) Yet another said, 'Who on earth is Dodo?' and Lord Ledgers patted the air appreciatively and said, 'You shall see.'

In fact, Lady Bretton's party, where all these opinions were put forward, was a great success, and every one recalled vividly the last time Dodo had been seen there. It was just after her baby's death—very shocking indeed; and then everybody sighed, and said that poor dear Dodo had been quite too brilliant that night. And when the dance was over, about thirty of Lady Bretton's more intimate friends stopped and talked about it till dawn. They all talked at once, and nobody listened to what anybody else said, and so they all enjoyed themselves very much.

Lord Chesterford was there, but nobody had the least hesitation in talking before him, for it was known that he had quite got over it. For himself, he did not talk much, but in one of the few pauses that occurred, he said something which was felt to have weight. It was this: 'It is no use guessing what Dodo will do, unless we guess so many different things that it only leaves her one course open. But we may certainly expect that the result will be perfectly unexpected. And as it has not occurred to any one that we shall all receive her with open arms, I venture to suggest that she will so manage—manage is too heavy-handed a word—that we do!'

Edith Arbuthnot—she had been married a year, and was strumming with her hands on the back of the sofa—sat up and looked at Jack.

'I shall never forgive her,' she said, with emphasis—'never.'

Bertie laughed.

'Oh, yes you will, Edith,' he said; 'you will even forget, not Dodo, but what she has done. You will end by writing another symphony to her.'

Edith turned on him severely.

'Bertie, the better I know you, the worse you know me. Since I married you I have gradually become a complete stranger to you.'

'Anyhow, I know you well enough to assert that you contradict yourself twice a day. Tomorrow you will probably tell me that you've asked Dodo to your party on the tenth.'

'And if I do,' said Edith vindictively, not caring to protest, 'you will be sure to ask me how I reconcile that with what I said before. Bertie seems to think,' she went on, addressing the company generally, with the air of a misunderstood martyr, 'that I am sent into the world to reconcile conflicting statements!'

'How very tiresome of him, dear!' said Lady Bretton.

'Especially when you have made the conflicting statements yourself,' remarked Jack.

Edith laughed.

'Any one can be consistent,' she said; 'and it must be very dull work. If you are consistent, you are one person only; if you are inconsistent, you are many. I am a whole houseful, from the scullery-maid upwards. Bertie says it's a constant source of excitement to him to see which of me is coming down in the morning.'

'You are all very late,' remarked Bertie.

Edith gasped.

'Considering that one of me bicycles before breakfast and another rides, I should like to know what meaning you attach to what you have just said.'

'I mean that on those mornings you order breakfast at half-past nine, and appear soon after ten.'

Edith rose.

'I am early enough now,' she said. 'The sun's just rising and I am going to bed, every one of me. If I was Dodo I should go for a ride. No, I shall never forgive her. Jack, we skate at Niagara remember, in the afternoon—the last skate we shall have till next year. Large Qs all round the rink. It's quite easy; but you must not make your turn as if you were a spluttering quill.'

The second bomb burst in the centre of All London next week, when, as has been already mentioned, Dodo replied 'Charmed' to the Duchess's invitation, and it was publicly announced that the bazaar would be opened on the second day by Princess Waldeneck. For the bazaar (fancy dress) was to be *the* smart thing of the season. No change was to be given, no silver taken, no untitled lady except Miss Anastasia M. Blobs, who was the rage just then—she could whistle through her fingers—was going to hold a stall, and there was to be a dance every evening. All London had suddenly realized that the greatest of all thing was charity, and it had pricked its fingers terribly over the discovery. Everybody brought little silk bags out to dinner with them, in which they kept their work, and after dinner sewed away diligently at squares of silk bedcovers and embroidered stoles. The effect on the thimble trade was perceptible, for every one kept losing their thimbles, and Mr Peter Robinson bought a house at Goring. Several young men even followed the fashion, and sat them chastely at their needlework in the manner of Penelope, and talked about rucking and tacking. In fact, the news that Dodo was going to open the bazaar on the second day made as great a sensation as if it had been announced that the angel Gabriel had kindly consented to do so, and the Duchess of Peterham, who was entirely responsible for it, was beset with questions. During the first week she had listened very attentively to what All London was saying, and she had drawn her own conclusions. She returned only one answer as the day drew near, but that bore repetition.

'My dear,' she said to everybody, 'we shall coin money like the Bank of England. I am told that certain people say they won't speak to her. Unless they come early they won't have a chance!'

The bazaar was opened in the time-honoured manner. The Princess had an enormous bouquet given her by an extremely small child, whom she kissed, and said she was flattered and gratified—she did say gratified and flattified, but that was not reported—to have been asked to fill this honourable and responsible post. The object of the bazaar, as usual, was nearer to her heart than any other object in the world. She had great pleasure in declaring it open, and walked round to all the stalls with a gentleman in attendance, who made a note of her purchases. Then she had some light refreshment, called tea because it was served at five o'clock, but consisting of little sandwiches and frills, and froth and sweets, and went away, promising to come the next day.

The Duchess turned round triumphantly on Lady Bretton when she heard this, and said, 'I told you it would be all right.'

Dodo arrived that afternoon from Paris with her husband, and they were met by the other secretaries and attachés of the Embassy. They dined at home alone; but it was known she had come, and opinions were divided as to whether she would go to the dance at the bazaar that evening or not. It was supposed to be 'early and late and large', and though by twelve the big hall was full, there was a general tendency to wait about in groups rather than to begin. The Duchess in vain tried to make people start, and the band diligently played Number One; but no one else took part in it, and conversation buzzed.

'Did you see her?'

'Only a glimpse. But Kodjek—he is the Second Secretary—don't you know him? Oh, there he is——Baron Kodjek, Lady Walling-ford—he tells me that she is perfectly marvellous, and not a day older. Isn't it so, Kodjek?'

'Did she say she was coming?'

'Waldeneck said he wouldn't go.'

'*Cela ne fait rien.* I imagine Dodo can take care of herself.'

'Who knows? The other day in Paris she wanted to——'

'All the Waldeneck diamonds? She wouldn't put them on just for dinner at home.'

'There's Number Two. I wonder if the Duchess knows.'

'Yes; she told me that in any case she wouldn't be here before one o'clock. But there's no reason why we shouldn't dance. If she comes, she comes. May I have the pleasure?'

Number Two was started, and by degrees the music exercised its legitimate function, and made dancing inevitable. The Duchess, who did not dance, went and stood on the platform near the door, and waited there. Number Three followed in due course, and was somewhat prolonged—in fact, it went on till a little after one. Then there was a pause—no one exactly knew why, but guessed; and the crowd on the platform near the door thickened. Everybody who was coming had come, but everybody crowded up towards the door, except a strongly marked but small contingent, who stopped at the far end of the hall, and wondered why the band did not begin Number Four.

Suddenly the door was thrown open, a Bayreuth hush settled down on the room, and one voice only was heard.

'It was too dear of you to get here just when I did. Jack, why didn't you come to dinner? Waldeneck would have been delighted to see you, and I, *ca va sans dire*. I really felt quite shy about making my entry here alone, and I believe I am not naturally so—no, Jack, it's no use telling me I am shy, for I am not; and now you shall take me in. Give me your arm. Ah, here's the Duchess! Dearest Gladys, it's an age since we met, and you've not grown a bit. Bertie, of all people, you! Where's Edith—oh, I never congratulated you about that; never mind, I do it now; and why isn't she here to play the Scherzo of the "Dodo" symphony on a comb, to welcome me. Am I very late? I suppose so—I always was; but Jack's just as late. Oh, come and dance somebody, at once. Yes, Jack, come along. Where have they got to? Number Four? Why it's a *pas de quatre*. My Serene Highness is in luck!'

The crowd on the platform made way for her and Jack, right and left, and she passed through, stopping now and then to greet an old acquaintance, chattering all the time, and stood for a moment under the brilliant electric light, on the edge of the dais, drawing on her gloves. She was dressed entirely in white, of some wonderful floating fabric that seemed to have been cut by the yard out of sea-foam and snow-storms. She had a great white feather fan in her hand, two rows of diamonds round her neck, and, set high in her black hair, a great diamond star. She had come back as beautiful, as devoid of all self-consciousness, as brilliant, and as different from all others, especially from those who had modelled themselves on Lady Chesterfield, as ever. And by her side, of all people in the world, stood the man whom she had used so vilely, whom she had betrayed and insulted in the sight of all the world. But for the moment he cared nothing; he only knew that Dodo's hand was on his arm, her voice in his ears, her eyes looking into his. His reason told him that he would pay for it afterwards, but he shrugged his shoulders at that, and refused to weigh 'afterwards' in the scale against the exquisite present moment; and, in the sight of all those who knew what had happened two years ago, he stepped down with her from the platform—and next moment the band had struck up, and they led the *pas de quatre*.

While they stood there, everybody except those to whom she spoke was silent, looking at her and wondering at her beauty and her incomparable charm; but as she danced—literally danced her way back into London again—tongues and feet were loosened, and in a couple of minutes there were a hundred couples following in the wake of Dodo and her partner. All London was as sheep, and Dodo was its shepherd.

Apparently she was as indefatigable and as fond of the *pas de quatre* as ever, for, with one short halt, she danced it through to the end, and then, still on Jack's arm, she went up to the platform again, and held a sort of drawing-room there.

'Jack, you must stop with me,' she said; 'I've a hundred things I want not to say to you; but I must speak to all these dear folk too. Hullo, Tommy'—this to Lord Ledgers—'you are a sight for sore eyes; not that mine are sore—thank you so much for asking: but, oh! you've taken to an eyeglass. Drop it, there's a dear. Yes, Madrid was horrible; but what was one to do? I feel like a hardy animal—or is it annual?—just bedded out again: do you bed out hardy animals? Really I am in luck. I only came out just to paddle in the water, as it were, and I seem to have taken a plunge into the middle of the deepest part. Isn't Edith here at all, Bertie? How tiresome of her! Oh, there's Maud. Dearest Maud, how are you? And how's Algy? Oh, Maud, you will never learn to be smart. You look like a badly bound church-service, with all those ribbons. They remind me exactly of book-markers. I apologize, darling. Oh, don't look so grave—or is it a saint's day? I shall come and see your baby tomorrow—I think it's simply cruel of you not to have called it after me. I suppose you thought it would be like giving a dog a bad name: you unnatural mother, to compare your baby to a dog! You don't like dogs, I remember—that makes all the difference; at least, I think I mean exactly the opposite, but I've no time to think. Jack, I've given up thinking lately. Congratulate me. Waldeneck's secretary does it all for me. He is all forehead, and when he thinks you can hear the wheels go click, click, click, inside his head. He's a sort of penny-in-the-slot machine. Oh, Jack, do you remember our getting butter-scotch at Bletchley Station and missing the train in consequence? Bletchley is a sort of General Confession. We did all the things we ought not to have done, like putting pennies on the line, and left undone all the things we ought to have done, like catching the train; and we didn't get any butterscotch either, because the thing stuck. Don't you remember? And if we hadn't said we were peers and peeresses we should have been taken up for tampering with the machine.'

'The first five pennies went all right,' said Jack; 'it was only the sixth that stuck. Your hands were full of butterscotch as it was!'

'Quite right—how well you remember!' said Dodo. 'We were wise to get a lot, because we had——dearest Evelyn, I'm charmed to see you. Jack and I are talking about butterscotch and Bletchley Station—it sounds like a tract, doesn't it? What was I saying, Jack? Oh, yes—we had a long journey before us, and we were young then, and boys together—at least, you were—and there was plenty of health in us. Gladys, are we going to have a cotillon? Do have a cotillon—and if you want some one to lead it, why, I'm your man. I haven't led a cotillon since—since last night in Paris. There was a dance at the Embassy, and one of the Greek princes was there, so those of us who had them wore the orders of Pericles and Aspasia, or something of the kind. Waldeneck was as cross as a bear because he had to go. Are bears very cross? The only ones I know really well are those at the Zoo, and they are angels without wings. Jack, where's my fan? It was made in Germany, but you'd never guess it. Oh, dear! how nice it is to be in London again! Every other country has a touch of "Made in Germany" about it. We had an accident coming from Paris to Calais: a coupling broke, and we nearly went off the line. They said the couplings were always breaking, because they were made in Germany. A stupid little French official told us about it and we couldn't get rid of him, so Waldeneck put his head out of the window, and said, "Alsace et Lorraine", and so he went away. That six months in Berlin added six years to my life, and tonight has taken eleven off, so I'm minus five. I don't look it, do I? I shall be a riot tomorrow and march down Piccadilly with a banner shouting "Give me back my eleven years!" Jack, come too—that will make twenty-two between us—we might have a cricket match. I want to dance again—no, not you, Jack; but the one after, if you like. Yes, come along, Tommy; you used to waltz respectably. I won't say more, or else you'll be conceited, and I hate a conceited man nearly as much as I hate a modest woman!'

Dodo could have danced all the way home, from sheer exultation. She had succeeded better than she dared to hope—she had taken London by storm. It had surrendered unconditionally, and she meant to take up her English life again exactly where it had stopped. She had directed her attack against what she knew would be the most impregnable part of London—namely, Jack—and in a moment her colours were triumphantly flying there. If Jack received her, she knew she was safe, for Jack's friends would, a priori, be her most bitter enemies. But if the head of the party goes over, his following, she reflected, would soon come too. And they had come, helter-skelter, after him. She had arranged to ride next morning at nine, and she felt quite certain that Jack would be there.

When she had gone the dance was over; but if she had stayed till nine next morning, it would not have been over till she went. Jack put her into her carriage, and then walked himself. It never for a moment occurred to him to consider what he had done. He had seen her. That was enough. The thing was over. He had been weak, foolish, culpable, and he shrugged his shoulders. The fact remained that Dodo was the one woman in all the world. The world would laugh at him—*soit*, he would laugh too. The world would call him objectionable names, and he would have tea with Dodo.

Waldeneck had gone to bed before she came home, and they met at breakfast. Dodo had enjoyed her ride immensely, and she came in very late, with her habit on, and a charming colour in her cheeks.

'You lazy old boy!' she said. 'You should have come last night, and also this morning. We had a delightful dance, and a delightful ride. London has taken me to its heart again, and I have taken it to mine. It is very pleasant for both of us. Jack said the season had been very dull—up till now, he was polite enough to add.'

'Jack?'

He spoke quite quietly, but there was something in his voice which Dodo had heard before, and had learned, not only to resent, but to be afraid of.

'John, Marquis of Chesterford, otherwise Jack,' she said, trying not to be nervous.

'Have you been riding with him?'

'He with me, I should say.'

Prince Waldeneck carefully wiped the froth of his coffee from his moustache before replying.

'It is positively indecent,' he said.

The colour faded out of Dodo's cheeks, leaving only two bright, angry spots.

'Is that all you have to say?' she asked.

'You can draw an inference from it, I imagine.'

'The only inference I can draw is that you are exceedingly rude,' said Dodo.

'Then I will point out another. It is this: I do not wish you to do it again.'

'I fail to see——' began Dodo.

'Pardon me. I have a word more to say. It is perfectly immaterial to me whether Lord Chesterford chooses to make an absurd fool of himself, but it matters considerably to me whether you do. We will have no more scandals, my dear Dodo. A scandal about his wife in my profession inevitably hurts a man's career, and I do not choose that my career should be hurt. And a woman—you may not perhaps agree with me—gains nothing by a scandal. She gets talked about, but she gets talked about in the wrong way and by second-rate people. I do not mean to quarrel with you for a moment. There is nothing so vulgar as a couple of people who quarrel with each other. I have heard you say so yourself on previous occasions. Get talked about as much as you like, but don't let it be in the wrong way. I wish to be quite reasonable, and you are clever enough to know what I mean, and also to be of immense use to me. Let us pull together, please.'

'I am not accustomed to be spoken to in the way you have chosen to speak to me,' she said. 'For instance, nobody in the world shall tell me that what I do is positively indecent.'

Waldeneck looked across the table at her, and for a moment regretted the expression he had used; for he knew she seldom got angry, but that when she did she became slightly unmanageable. So, though it was not his way to withdraw either words or actions, thinking to secure his point more easily, he changed his tone somewhat.

'It is the effect of your English climate,' he said. 'In London people speak with a frankness which is quite brutal. I daresay I spoke too frankly. There is no manner of use in being angry, Dodo. Come, I wish to be reasonable. Be reasonable too. Take a warning that is meant to be in your interests no less than in mine. Remember it is possible even for you—and I do full justice to your charm—to go too far. The world had forgiven you a good deal because you are clever and brilliant, and because you amuse it. For that very reason don't strain its powers of forgiveness till they snap.'

Dodo rose from the table, and walked across to the window to pull up the blind. This was all she wanted; for all that she feared in her husband was his power of command: as soon as he came down to advice and persuasion, she knew from experience that it was because he was not certain of his own mastery. If he had stopped when he said, 'I do not wish you to do it again,' she would have sat tongue-tied and beaten; but he had condescended to make an appeal to her reason.

She broke out into a perfectly natural laugh. 'No, we won't quarrel,' she said, 'or we should be like the two magpies in the nursery rhyme, of which nothing was left but the beak of one little magpie, and the other little magpie's tail. But we won't. I intend to be perfectly reasonable—only recollect that, and be easy in your diplomatic old mind. By the way, Gladys was there last night, in all her glory. Don't you want to renew your acquaintance with her?'

Prince Waldeneck felt that he had not quite gained his point. Dodo's admirable promptness in regaining her temper made it difficult for him to revert to the subject, and her allusion to the Duchess was particularly disarming, for he had, a year ago, flirted with her in a way which Dodo might have resented if she had felt so disposed. But she had laughed at it all then, and told her husband all that people said to her; and, with infinite tact, she had chosen this moment to laugh at it again. So he merely replied—

'I shall be delighted to. You are opening the bazaar today, are you not? I will come with you, I think.'

This possibility was distinctly not on Dodo's programme, but she seemed to accept the alteration with alacrity, and finished breakfast.

If Dodo had been a success the night before, she can only be described as a creation that afternoon. Waldeneck was detained at the Embassy, and she went alone, dressed as a deified Albanian peasant. Every one who had been there the night before, and every one else who had not, rose at her like one man. She had disgraced herself, she had dragged Jack's name in the dust, she had been execrated and vilified, and she had even, which is more fatal, been half forgotten; but as soon as she appeared it was seen that she held the unique distinction of being herself. The Royal Princess was at an enormous discount; ladies curtsied to her and looked over their shoulders to see what Dodo was doing, and gentlemen hurried away from her side to go to Dodo's improvised stall, which consisted of mixed drinks, photographs of celebrities, buttonholes, boot-jacks, palmistry, and the reading of character from handwriting. Even Miss Anastasia M. Blobs might have whistled through her fingers till she burst and no one would have congratulated her. And Dodo did it all: she mixed the drinks; she signed the photographs with any name that occurred to her, usually her own; she pinned the buttonholes; she sold the boot-jacks; she confused the line of heart with the line of life, and, for aught she knew, the line of head with the line of feet; she told the most elaborate characters—an art of which she had no previous knowledge whatever; and she raked in gold.

Later on her husband came, and joined in the crowd which surrounded her stall. She seized on him at once, mixed him whisky-and-soda, fastened a large camelia in his buttonhole, gave him a photograph of Henry Irving, signed, 'Your sincere friend, Dodo', and, pouncing on his hand, opened it and gazed at it with a gesture of dramatic despair.

'Ladies and gentlemen,' she said—— 'Oh, I beg your Royal Highness's pardon; I didn't see you'—Dodo curtsied, but continued without a pause—'we have before us one of the most battered pages in what I may call the Book of Doom. This Englishman—the hand is unmistakeably English—is of miserable and dwarf-like physique, being probably not more than five foot two in height.' (Waldeneck was six foot three.) 'He has never known the softening influences of married life, and his career, if we may call it a career, has been one of misfortune and calamity, for which, as Mr Hawkins melodiously sings, he has only himself to blame. His want of affection is only equalled by his lack of intellect, and I am sorry to observe that the line of morality, like that of political economy, is entirely absent. He will be burned at the age of forty-seven—let's see, you are forty-seven, aren't you, Waldeneck—I beg your pardon, thirty-seven, and in a garret in the New Cut by the hands of the common hangman. There! you've had a buttonhole, a whisky-and-soda, a boot-jack—did you have a boot-jack, darling?—blow the expense; have it now—a photograph of our eminent tragedian—or was it a photograph of Mr Stokes of the Congo?—and a fortune. Reduction on taking a quantity—no, I don't think any reduction—five pounds please, and think of all the poor children who will be glad of it.'

Dodo wrote an IOU, signed it with her husband's name, and put it into her bag.

'There's nothing like making yourself pleasant, Jack,' she said. 'You haven't got a buttonhole. Do you think a cabbage rose would suit your complaint, or would you fancy this?—I can't bear people who say "fancy". It looks to me like a potato-flower, which is said to be deadly poison. Don't eat it now, or you'll have to buy another. Yes, I think it's uncommonly cheap. I'm rather cheap, too, today; but you see I've danced for the last two nights, and the two nights before I was in the pull-puff. Oh, Jack, England's a good place. Thank goodness I remain an Englishwoman. How I shall enjoy my nice white bed tonight! My bed will be like a little boat-hook, as Stevenson says, when I ought to get up tomorrow. What'll you take, Tommy? There's gingerine, or portine, or something, for tee-totallers—all temperance drinks end in "ine"; and no one is allowed to be drunk on the premises. Why didn't I think of selling cigarettes too? I would give my immortal coil—I mean my soul—for a cigarette this minute—no, thanks, I daren't; they'd be shocked—or should I have had to get a licence? "Dodo, licensed to sell tobacco and boot-jacks." I might have sold the licence too. Waldeneck——oh, has he gone? Go and look for him, there's a good boy—and tell him that I am dining out; and when you come back I'll give you some claretine gratis. Yes, one sovereign please—no, I always take the money first, if it's all the same to you. Yes, this is evidently a woman's hand,' said Dodo, seizing hold of Bertie, 'and it presents some curious features.'

But Waldeneck had disappeared, and Dodo did not see him again till, coming home late that night after the dance, with an armful of cotillon toys, she found him waiting up for her. His eyes looked rather red, and he staggered slightly as he walked across the room to the door. He had clearly been simply waiting up for her to come home. His voice, when he spoke, sounded thick.

'So you have come back,' he said. 'Cannot you see that all those people were laughing at you? Is it possible that you do not know that you are growing too old for that sort of foolery? And the way you spoke to me, telling me all that stupid gibberish, was an infernal impertinence.'

Dodo stared at him in disgust and amazement, and all her toys fell rustling to the ground. She despised him and shuddered at him in that moment, and she spoke quite collectedly, in pure contempt.

'You had better go to bed,' she said. 'Some of the servants may see you, and, though you may not know it, you are not fit to be seen.

Besides'—the sarcasm was irresistible—'does it not hurt your career
to drink too much?'

DODO'S PROGRESS

M ISS GRANTHAM had secured what she called 'five minutes
edgeways' with Dodo on the second night of the bazaar. This
meant that Dodo talked to six people at once, of whom she was one,
firing sentences off at each in turn with moderate regularity and
inconceivable rapidity. Her minute-guns to Miss Grantham formed a
fairly coherent whole, and were something to this effect.

'Of course I sha'n't allow Edith to *bouder*; it is quite intolerable
that she should do that. I think she must have got what they call a
nonconformist conscience. I mean, her conscience won't let her
conform to what everybody else conforms to—me, for instance. Yes,
Grantie, I know she's a perfect darling, but perfect darlings have
always something rather queer and cornery about them—how do you
call them? Polygonal figure, isn't it—Edith hasn't a very good figure.
She's like one of those india-rubber faces, which you squeeze
together—all features, and not enough room for them. She tucks her
music under one arm, gets on her bicycle, and slides over Bertie, if
you know what I mean—and, oh, do you know the way her eyes
become like large, bright buttons when she is busy nonconforming? I
shall certainly go to see her tomorrow morning: no, I'm not the least
afraid, thanks, and shall talk to her quite beautifully, like a sheep that
was lost. Are you going there too? Then tell her that she may expect
me, and that it will be not the slightest good her saying she is out.
Besides, I make an appointment with her, so she must keep it. I shall
allow her to be a sort of moral dentist to me, and tell me that my
morals want stopping.'

Accordingly, next morning Grantie went to see Edith, whom she
found in her private and particular room. Edith had insisted on an

unwritten agreement between herself and Bertie when they married, that she was to be allowed a room where no one might come unless she wished. Bertie had agreed, on condition that he might have one too, and the house was consequently divided into his *chez-moi* and her *chez-moi*, and neutral or common ground on which they entertained their friends. Edith kept Miss Grantham waiting, on principle, for a few minutes, at the end of which she was admitted, to find Edith scoring the music of a symphony. The table at which she worked had grown pyramidal in shape, owing to an accumulation of various materials on it, and Edith worked about half way up. A bicycle stood in one corner of the room, and a book-case, with the complete Badminton series in another. The window was wide open, and an exceedingly cold draught whistled round the room, occasionally fluttering the outlines of the pyramid. Edith did not look up as Grantie entered.

'Good-morning, Edith,' said Grantie politely.

'Morning. What do you want?'

Grantie walked across to the window, shut it, and sat down gracefully on the window-seat.

'I wanted to see you—pure affection: no less, and no more,' she said, 'though you do put your friends a little lower than your bicycle, and a great deal lower than your music.'

Edith did not reply, except to murmur, 'Then the flutes take it up,' and Grantie went on, in a slow, persuasive voice.

'What was I saying? Oh, yes, Dodo was a great success last night. She came to the ball, you know. Really she *is* successful. It must be so nice to be sure that every one who knows you is talking about you, and that every one who doesn't is doing the same, and pretending they do. Really the proper study of mankind is Dodo. Now I can't help wondering——'

Edith laid down her pen.

'Grantie, have you come here just to talk about Dodo?' she asked.

'Oh, no: I have a lot to say to you. Oh, by the way, Dodo said that I was to tell you to expect her this morning, and that it was no use your pretending to be out. Has she been yet? I'm rather late.'

Edith frowned, and rang the bell.

'If any one calls,' she said to the man, 'say I am out. Any one, mind.'

Miss Grantham was a little tired with her dance the night before, and only sighed gently; but, realizing that the meeting between the two might be interesting, resigned herself to be battered by Edith in monosyllables till Dodo appeared. She had taken up an old number of a magazine, and laid it open, among the advertisements, upside down, on her lap. Edith, who was rather vexed at having been interrupted at all, and who was boiling with indignation at the thought of Dodo's further intrusion, waited for her to speak, in order to have the opportunity of contradicting her. But after a minute or two it appeared equally possible to contradict the meaning of Grantie's silence.

'You seem to think,' she said, 'that I am going to do as Dodo tells me, and just because other people are delighted to see her back, that I too shall welcome her. Grantie,' she continued, with growing emphasis, violently dotting a crotchet, as if she were stabbing it, 'how little you know me!'

'You must remember that I never said anything of the sort,' said Grantie mildly, but facing round a little.

'No, but you implied it,' said Edith.

Grantie, knowing in her own mind that she was willing to lay odds on Dodo, was silent a minute. 'It would be as good as—as a theatrical representation, to see you and Dodo meet,' she said at last.

'I suppose you mean a play?' remarked Edith, and took up her pen again.

But Grantie did not notice this last shot at her; she had seen Dodo's carriage draw up at the house from her seat in the window, and she waited with immense interest for developments. She found herself regarding the upside-down magazine in the absent way one looks at a programme just before the curtain at a theatrical representation goes up.

The first development was the footman, who had been told to say that Edith was out. He seemed a little flustered and nervous.

'The Princess Waldeneck to see you 'm,' he said.

Edith looked up, and, in case Dodo was on the landing,

'I told you to say I was out,' she said, with great distinctness.

'Yes 'm, I did. But the Princess told me she thought not.'

Any further discussion was saved by the entrance of Dodo herself.

'I was told you were out, dear,' she said, 'but I thought it safer to see for myself, because sometimes that means so little, and is so misleading. Oh, I know the Englishwoman's house is her castle, but I really wanted to see you very much. What a nice castle! Hullo, Grantie, the top of the morning to you. Grantie, dear, will you please go away, and sit in the servants' hall or something, till I've finished talking to Edith?'

'I'm not at all sure that you are going to talk to me,' said Edith, as stiff as a poker.

'Dear Edith, I have come here to convince you of it,' said Dodo. 'Now please go, Grantie.'

Grantie rose gracefully and unwillingly, and Dodo shut the door behind her. Then there was silence, Dodo being busy arranging her thoughts, Edith stiff and truculent, but, despite herself, a little interested. But this she put away as a weakness unworthy of her.

'As you may have heard,' said Dodo. 'I have come back. Candidly, I think people are glad to see me.'

'So I have heard,' said Edith, simply incapable, now Dodo had forced an entrance, of not treating her to a little candour. 'I am told that you whistled to all your old friends, and they came running after you like sheep.'

'I never whistled to a sheep,' remarked Dodo parenthetically.

'But you may whistle till you burst,' continued Edith, with vehemence, 'before I come running after you.'

Dodo sighed, and drew off her gloves.

'Dear Edith, I never meant to whistle for you, for I always have known that you are not one with whom whistling succeeds. Please give me a cigarette—I can talk more with a cigarette, though I can talk a little without. I knew I should have to see you, and explain things.'

Edith laughed.

'Thanks, I do not want things explained to me,' she said. 'I am perfectly satisfied that I have a very tolerably clear idea of what happened. In fact, it was not a private matter. And now,' she concluded, 'I am going into another room. When you are tired of stopping here, will you please ring the bell? and the man will show you out.'

Dodo walked quietly to the door, locked it, put the key in her pocket, and sat down.

'Now, Edith, don't be angry,' she said. 'I am determined to talk to you, and you may be certain I shall. It's no use your ringing the bell, for they will have to send for a blacksmith, or a locksmith, or whoever does that sort of thing, and by the time he is here, I shall have finished.'

Edith stared at her a moment.

'I think you are mad,' she said, and sat down again to her work.

Dodo was silent a moment longer, deliberating intently with herself, and then broke out into speech.

'How can you treat me like this?' she said. 'How dare you behave in such a narrow and conventional manner? Really, Edith, you are surely learning to be an understudy for Mrs Grundy. Not that I object to respectability, in the least—in fact, I respect it very much; but you are presuming to judge me without knowing a single thing of what has happened. It is Mrs Grundy's infallibility on questions of which she is ignorant that offends me in her and you.'

Edith again felt a little interested. After all, Dodo was not often tedious, and why not have a quarter of an hour's talk with her, which might be entertaining, and which would certainly fail to convince her? But for the present she merely drummed with her hands on the table, and said, not to Dodo, but to the ambient air—

'I was never called Mrs Grundy before.'

'How you can sit there and pass judgement on me, when you know nothing of what has happened—nothing from the inside, at any rate—is incomprehensible!' said Dodo. 'Have you no better notion of friendship, or loyalty, than that?'

'I don't pass judgement on you,' said Edith. 'I am completely indifferent as to what you do. No doubt we shall often meet, in other houses. I shall not leave the room when you enter it, and I daresay I shall sit and talk to you as friends talk——'

'Friends!' interrupted Dodo. 'Heaven help us! Are the indifferent people one meets like that, friends? Is that your idea of friendship? Were we only friends like that? And if more, what do you suppose loyalty means?'

'Then I will put it differently,' said Edith. 'And if you find my words unpleasant, you must remember that I did not insist on this interview.'

'No, dear, I confess you didn't insist on it,' remarked Dodo, with the ghost of a smile.

'What I have to say is this,' continued Edith inexorably. 'I do not

care for the society of a woman who has behaved as you behaved to
Jack. Is that your idea of loyalty?'

'Jack has forgiven me,' said Dodo.

'Because one person is a weak fool,' remarked Edith, 'there is no
moral obligation for me to be a weak fool too.'

'Poor, dear Jack!' murmured Dodo. 'Let us leave him out.'

'You seem to leave him out pretty completely,' said Edith.

Dodo clasped her hands together.

'Edith, don't speak to me like that,' she said. 'Sharp words never
mended a matter.'

'I do not wish to mend this matter,' she said. 'It is irreparable. You
seem to think your friends are like gloves: you can put them on, and
then pull them off and throw them down if you like, to pick them up
again afterwards, if it suits you to choose to do so.'

Dodo turned on her sharply.

'Ah, you are unfair, you are atrociously unfair!' she cried. 'When
did I ever cast off a friend? Is it like me at all to drop people who
have been my friends?'

Edith hesitated a moment.

'No: I am sorry I said that,' she replied. 'It is not like you.'

A tap came at the door, and Miss Grantham's voice, with a
plaintive patience, demanded admittance.

'You've been ever so long,' she said. 'Mayn't I come in?'

Edith hesitated.

'No, Grantie,' she replied at length. 'Go away, please.'

'Thank you,' said Dodo, looking up at Edith.

Then, after a pause, and with a more rapid and rather tremulous
voice:

'You see how it is, Edith,' she said. 'You make hasty judgements in
your own mind about me like that, and unless I am here to
contradict you, you endorse them and docket them, and put them
away in your mind as things proved and demonstrated. And when my
name comes up, you look at the Dodo drawer, and find I am disloyal
and drop my friends. You are horribly unfair; and about the great
matter, on which you are unfairest of all, you know nothing,
absolutely nothing. You have judged hastily and superficially. Just
because you are strong and self-sufficient yourself, and love your
music, and Bertie, and bicycling, you think all others have an equal
abundance of soul-filling material at command. Because you yourself

love Bertie in a comfortable roast-beef slap-on-the-back way, you think there is no other kind of love. You do not know that a woman can be carried off her feet, whirled away, impotently. To be swept away in spite of one's self, best self, worst self alike; to be—Ah! what is the good of my talking to you? You will not understand, and I don't think you want to.'

Edith was distinctly interested. She had listened to Dodo at first with the air of an *entre* at a play, and a certain silent applause she gave her was given as by a critic to an actor. Surely she was acting. Yet why should she act to so meagre an audience? And if she was not acting, it was certainly a reason the more for listening to her. By degrees the probability of its genuineness grew on her. Dodo, she could not help feeling, at any rate thought herself in earnest; and when, at the end of her speech, she looked at her for a moment with hopeless appeal in her eyes, Edith was touched. And Edith was right. Dodo was perfectly in earnest, and though she put the truth in the most effective way she could think of, it was the truth for all that, though most carefully dressed up, and entirely fit for public appearance. Edith got up from her chair, and sat herself down in one closer to Dodo.

'Try to explain it to me, Dodo,' she said. 'I will listen to you, which at first I was not willing to do. But now I think you care that I should.'

Dodo made a mental note that at last Edith had called her by her name, and went on.

'It is simply what I say,' she continued. 'I was blind, deaf, dumb: I could not, I was morally incapable of resisting. If Waldeneck had told me to throw myself out of the window instead of coming to Paris with him, I should have done it. I might have begged for a minute to put on my hat, but I should have done it. He is strong—good heavens! he is strong. You don't understand what that means—to find some one stronger than yourself, and who can beat down resistance as an iron bar can beat down a weaker thing.'

'No, I don't,' said Edith. 'I believe one has free will always and continually; one need do nothing unless one chooses.'

Dodo stared silently before her, smoothing out the creases in her gloves.

'That is a very comfortable idea,' she said; 'and, like most comfortable ideas, perfectly impracticable. Free will ceases the

moment you meet somebody stronger than yourself; and I——'
Again she stopped, her voice trembling.

'What is the matter, Dodo?' said Edith. 'I don't in the least
understand why you should have come here. Surely you have enough
friends! Are you not happy?'

'Happy?' Dodo burst into a mirthless laugh, and then, without a
sign or a word of warning, buried her face in her hands and burst
into tears—hopeless, desolating sobs, like a child over a broken toy.
At that Edith melted altogether, and jumped up with a face full of
helpless concern.

'Oh, Dodo, what can I do?' she asked. 'Shall I—shall I play to you,
or will you have a cigarette? Only don't cry like that.'

Edith ransacked her brain to remember how people behaved to
her when she cried, but she could recollect no instance of her ever
having done so. Meantime, Dodo, with an absolute disregard of lace
and gloves and hat, sat crushed together in her chair, abandoning
herself to what was partly a relief, but wholly natural: and Edith
offered her cigarettes and a smelling-bottle, and walked about
agitatedly.

Her fit stopped as suddenly as it had begun, and she sat up, put
her hat straight, and blew her nose. 'There, I am better,' she said:
'and we will go on talking. Happy! Oh, Edith, you don't know that
man. He is a brute, I tell you—a brute. He offends me, and grates
on me. I believe I am not naturally prudish,' she said, with a slight
return to her normal manner, 'but he shocks me. I came home last
night. He was sitting up for me; he was drunk, he was beastly drunk,
and he said abominable things to me. Thank God, I am less afraid of
him now. At first he paralysed me. I could do nothing but what he
told me; but now, though he gets his own way, he doesn't always get
it in the way he likes best. I said to him last night what I shouldn't
have dared to say a year ago. I——'

'Oh, stop, stop,' cried Edith. 'He is your husband. Can you not be
loyal to any one?'

'Yes, I know he is my husband,' said Dodo, 'but that is his
concern, not mine. It was his doing and no other's. Loyal! Is an
animal loyal to its trainer. Are lions loyal to those hideous women in
tights who crack whips at them? Why, they are afraid, that is all; and
if they cease to be afraid—well, you have a sensational little
paragraph in the papers next day. He made me afraid—I married

him for that reason; but now I am less afraid. Oh, there will be no sensational paragraphs—there is nothing so vulgar as washing your dirty linen in public. But never mind about that. I came to be friends with you, and I have told you what I should only tell to a great friend. And I have cried, and I feel better: I do not often cry, and I never do it on purpose.'

Edith laughed.

'Oh, Dodo, really I didn't suspect you of it!'

Dodo sat up and continued smoothing her ruffled plumes.

'Are you sure? Are you quite sure?' she asked. 'For indeed, there was no enormity you would not have been willing to believe of me when I came in. Oh, there's Grantie knocking again. Let her in, Edith. Where's the key-catch?'

Miss Grantham came in, looking a little ill-used, having been excluded from the performance; but curiosity conquered pride, and she looked at Dodo with raised eyebrows and intense expectation.

'Yes, we've had a very nice time, haven't we, Edith?' she said in answer, 'and we are all a happy family again, and I'm the white elephant. We talked about free will, and we both agreed—at least, now I come to think of it, we didn't; and we told each other a quantity of secrets, and settled that two was company, and three was more company, and so we let you in. Oh, we're going to have quite a lot of company on Friday next—of course you will come, Edith? Bring Bertie. Come to dinner too—no, on the whole, don't, because I think we haven't got enough finger-glasses. It is the Queen's birthday, and as Count Vramhoff is away, Waldeneck has to make the speech; he says he's sure that the Queen has at least two birthdays a year, and he hates making speeches, and he says nobody enjoys listening to them. But then, he is naturally modest.'

'Modesty is a pose,' said Miss Grantham, with an air of finality. 'Of course, everybody poses, and it is all right enough; and the really commonest pose is to pose as being natural.'

Dodo drew on her gloves, with her head a little on one side.

'No, you're wrong,' she said. 'I don't pose: I don't think Edith poses. Yes, I'm sure Edith doesn't pose.'

'You both do. You both have the natural pose.'

Edith gasped.

'Grantie, you have said the silliest thing I have ever heard,' she remarked.

'Well, I gave you credit for posing,' remarked Grantie. 'It is not decent not to pose, and I don't know what other pose you have. It is like having no gloves on.'

Dodo laughed.

'Dear Grantie, don't be so natural,' she said. 'Come and drive with me in my carriage. I am going to go round by the park, and home for lunch. You shall have lunch too and see the world-wide performing Austrian menagerie. We all do our tricks, and Waldeneck stands in the middle with a whip, like that man at the circus—how do you call him?—the—the *écuyer*—and sees we do them perfectly.'

Grantie looked up.

'I think that sounds interesting,' she said. 'Yes, I will come. Do you usually do your tricks well?'

'Admirably. My trick is to prevent people talking to him, or rather to save him talking to "other brutes", as he says. Consequently I am at my best. Yes, modesty is certainly a pose, and I haven't got it. Well, goodbye, Edith; and you will come to see me, and I will come to see you, and the world will wag its tail at us generally. Just now a section of it will soon want to be fed, so I must go. It has been charming to see you and to talk to you.'

Dodo chattered her way out, and she and Miss Grantham got into the victoria and drove northwards. Edith remained a full minute looking out of the window in thought, and then went to find Bertie, whom she always consulted when she wanted some one to agree with her.

'Bertie,' she said, 'Dodo has been to see me.'

'And you have made it up, and have come to tell me so,' said Bertie perspicaciously.

'Yes; and I wish to tell you also that it is very stupid of you to think me inconsistent.'

'Well, I won't ask for your explanation of this seeming inconsistency,' said Bertie amiably. 'I will take it for granted that it is really the truest consistency. Will that do?'

'No. We have both misjudged Dodo. We were not inconsistent, but wrong.'

'I never judged Dodo at all,' said Bertie; 'therefore I can't have misjudged her.'

'Then you ought to have judged her,' said Edith, 'and not credenced that sort of thing. Oh, really you are rather trying.'

'I know—I mean to be. You have invaded my *chez-moi*, and it is against the rules.'

'Well, I'm going to Dodo's on Friday.'

'All right,' said Bertie severely.

'Aren't you surprised?'

'Not in the least.'

'Why not?' she asked.

'Because it is part of my profession never to be surprised at you. I am working hard at it.'

Edith laughed.

'Well, that's all right then. But Bertie, I should prefer for the future that you should be surprised at me oftener, and then always be immediately convinced by the reasons I give you.'

'I will make an effort,' said Bertie; 'but it will require an effort.'

Prince Waldeneck, meantime, had spent a solitary and thoughtful morning. He recollected, but only vaguely, the event of the night before. Dodo, he remembered, was out late, at that stupid bazaar. He had waited up for her; he had drunk rather freely; she had noticed it, and, what was more, had told him so with some frankness. He knew, and had known for a long time, that she was an exceedingly useful person to him, or could be so when she chose—or rather, as he said to himself, when he chose. Once she had smoothed over a diplomatic affair, which promised to be disagreeable, almost impracticable. He had taken her into his confidence completely—at any rate, completely enough to convince her that she was possessed of the whole matter—and had told her frankly that she could help him in a way that no other person could. At the time he was in charge at Madrid the Italian chargé d'affaires was inclined to show himself obstinate about an international matter which his government had been prepared to face and make the best of—a rather serious difficulty. His Italian colleague was an amorous old fellow, with the firm conviction that he was in love with all agreeable women, and that all agreeable women were in love with him. Dodo, so he thought, was an agreeable woman.

Waldeneck had gone to Dodo's room that morning, prepared to be charming to her. He explained to her the whole position. A great deal depended on the personal feelings of the Italian; he would make the matter strike his government as it struck him. By nature, as Dodo knew, he was a silly old man. Would she turn the disposition of this silly old man to good account? And he read her a letter from the Austrian foreign secretary.

Dodo considered a moment.

'He is to make love to me, you mean,' she said, with almost embarrassing frankness, 'and immediately afterwards to talk business with you.'

Waldeneck laughed.

'Yes, I mean exactly that,' he said. 'You can always cut him afterwards.'

Dodo nodded.

'The great point in being married,' she said, 'is absolute frankness, and it deserves to be rewarded. I will make him tremulous. I will cut him afterwards, because he bores me.'

She had succeeded to admiration. Waldeneck found great difficulty in making him talk about business at all, so loud was he in the praises of this inimitable wife. And now Dodo, this eminently attractive and useful person, had presumed to tell him, her husband, that he was drunk and not fit to be seen by the servants. No diplomatic relation was served by this.

Dodo returned from her drive, and met Waldeneck in the hall. She had spoken her mind the night before, and had no intention of opening the subject again. He spared her the trouble.

'You chose to speak to me last night,' he said, coming close to her, 'in a way that no one speaks to me. Please remember that. How do you do, Miss Grantham? My wife, I am charmed to see, has brought you to lunch with us.'

The colour faded from Dodo's cheeks. Miss Grantham, who stood close to her must, she knew, have heard Waldeneck's words. She turned to her quickly.

'Go upstairs, Grantie,' she said. 'I will come in a moment. No, I will come with you now.'

Then, raising her voice a little—

'Dear Waldeneck,' she said, 'you speak English so well, but you will never learn to be really English. In England we never wash our dirty linen in public—even in the most strictly limited public. Be ready for lunch, old boy, won't you. Rodjek is coming, and every minute with him before lunch is ten years off my life; and he is always punctual.'

She passed upstairs with her arm in Miss Grantham's, and the latter could feel it trembling. Dodo shut the door behind them, and threw herself into a chair.

'Grantie, if you ever breathe a word of what you have heard, I will cut your throat,' she said.

Grantie's eyes were wide with interest.

'Oh, but it is so interesting, Dodo,' she said. 'Won't you tell me all about it?'

'Not another word.'

Waldeneck stopped in the hall, watching them ascend. And a sense of imperfect mastery gnawed at him. He felt uncomfortably sure that a year ago Dodo would not have been able to say that to him.

SPOOK
STORIES

NUMBER 12

IN the course of many shiftings and pitchings of my tent in this delectable town of London there has always been one particular square which I longed to live in, and in that square one particular house. Several times, too—but not often, for a house in that square is rarely publicly advertised as vacant—I could, in the course of the last fifteen years, have obtained the half of my desire and got a house in the square, but on each of these occasions I have felt that it would have been more tantalizing than I could bear to be so close to the house I wanted and yet not be in it. I should see it every day, I should pass and re-pass it, and it was better to be far away and out of sight of it than to have my desire continually whetted by its inaccessible propinquity. So I just continued wanting it, and since there is nothing in the world more true than that if you want a thing enough you get it, it happened that last summer I got what I wanted. I got, in fact, rather more than I wanted.

But before I got it there came one most dreadful disappointment. For fifteen years, as I have said, this particular house was never vacant, but one evening when I passed through the square on my way home I saw what I knew I should sometime see, namely, a board up, on which it was distinctly stated that No. 12 was to be let unfurnished. The name of the agent in whose hands it was appeared there, and I went straight to his office; but business hours were over, and it was shut. I rang him up on the telephone early next morning, only to find that No. 12 had been taken half an hour before. When next I went through the square there were furniture vans at the door. But this catastrophe neither diminished my desire for the house nor a sort of lurking certainty that I should get it, and I continued wanting it quietly and insistently.

Then occurred a small coincidence. I took into dinner a week or two afterwards a woman whom I had not previously met, and I am afraid we violated the rules that govern conversation at that meal, for we did not when half-way through turn our heads right and left, in place of left and right, and talk to our neighbours, but continued our animated discussion until the end. The subject was ghosts and supernatural phenomena generally, and Mrs Aylwin had a plausible and interesting theory on the subject of haunted houses.

'I can't believe,' she said, 'that if you or I had the misfortune to be murdered our spirits would be saddled with the further misfortune of being compelled to hang about the place where we were done to death. It would be too grossly unfair.'

'But it isn't usually the murdered who haunt the place,' said I, 'but the murderer. Mightn't that be a form of expiation? And then there are suicides. . . . There are a lot of apparently well-authenticated phantasms of suicides. That might be expiation, again.'

I was looking at her as I spoke, and for a moment thought I saw an expression of terror come into her face, and wondered whether she had some experience of the sort. But it instantly passed.

'Oh, I can't agree with that at all,' she said. 'Fancy having to go on committing your crime year in, year out!'

'What is your theory, then?' I asked.

She sat silent a few seconds.

'Put it like this,' she said at length. 'If some crime of violence has been committed in a room there must have been a terrific outpouring of emotion, hatred, terror, despair, or what not. I can imagine that it sets up certain vibrations, atmospheric, etheric if you like, and that when someone who is sensitive to such things comes there he may be, like some wireless apparatus which catches and records these vibrations. . . . Ah, my eye is being caught.'

We did not renew our conversation afterwards, for there was bridge to play, but when she and her husband left she asked if she could give me a lift. My house was on the way, it appeared, to hers, which was No. 12 in the square I have spoken of.

I was out of town for the next fortnight, and having been invited to go and see Mrs Aylwin some afternoon, I took, on my return, the earliest possible opportunity of doing so. I found the house shut and shuttered, and a board up to say that it was to let. On this occasion I was in time, and before the week was out my lease was signed. The price asked for it, considering the excellent repair the house was in, was most moderate, and by the middle of July I was installed.

The feature of the house, as I had conjectured from external scrutiny, was the long panelled room on the first floor, the five windows of which looked into the square. But I had not foreseen the admirable Adam fireplace that stood in the centre of the long wall, nor guessed the square hall below, nor yet the dining-room built out at the back of the house and overlooking Regent's Park. Below it was the kitchen and a tiled yard that communicated with a back gate. Here, too, well out of sight of the dining-room, was coal-cellar and wine-cellar. All was perfectly planned, and now that I had got my desire no leanness of the soul followed, but I lived as in an enchanted dream, and began seriously to consider whether I should not stop in London through August just for the pleasure of living here.

The weather broke in the last week of July, and I came home one dark and drizzly evening, congratulating myself that I was not dining out, but would enjoy a solitary and uninterrupted seance with odd jobs and tidyings that were still unaccomplished. One of these was to label and ticket the racks in the wine-cellar, and on getting home I went out across the tiled yard at the back to effect this. When I had finished it was already nearly dark, but the lights shining out of the frosted window of the kitchen illuminated the paved yard.

And then I saw what made my heart for a moment stand still, for in front of me there imprinted themselves on the wet tiles the mark of footsteps preceding mine, as if there was some invisible presence there that manifested itself by the impressions of its footfalls. They came from the coal-cellar next the scene of my labours, preceded me to the back door of the house, and, more visible now on the dry flags of the passage within, went to the door of the butler's pantry. There they stopped, and, going inside, I found the pantry empty. When I came out again the footmarks had already faded. But though the pantry was empty I had the feeling that there was someone there.

I went upstairs to dress for dinner, delaying for a moment in the hall over the post; the first slight pang of fright had passed, and I found myself only immensely interested. Inexplicable as the thing was, it was yet so silly. . . . On my way to my bedroom I just looked in at my beautiful long room, where the curtains were still undrawn, and saw the figure, I supposed, of my man-servant brushing up the hearth of the Adam fireplace. He was busy at it, and did not see my head put round the door, and I went on my way. Half-way through dressing it suddenly occurred to me to ask myself why he should be making the hearth tidy when there had been no fire there. And simultaneously I heard a step coming down from the floor above, not up from the floor below, which sounded like his.

As it passed my door I called to him, and he entered. 'Were you in the drawing-room just now, Sanders?' I asked.

'No, sir,' said he.

And yet five minutes ago I had seen with my own eyes the figure of a man there. . . .

These two occurrences were inexplicable enough, but for some reason they did not much disturb me. They had the texture of dream and unreality about them: they were, in spite of their oddity, dim and without vividness, and I sat that evening in the long drawing-room, where I had seen the figure, in a perfectly tranquil frame of mind, first reading the evening paper and then, about ten o'clock, settling down for an hour's work. I sat at a table in one of the windows, and, quite absorbed in what I was doing, let the time pass unheeded.

Then suddenly there came into my brain, with the sharpness of an order received, the command that I should turn round and look at the fireplace. . . . There in front of it, with his back to me, stood the figure of a man: his arm moved as if he was sweeping ashes together. . . . Then he turned, and with noiseless tread went towards the door. He was in profile now, and I could see his face. Then he was there no more, but the door had not opened to give him egress.

For the next week I saw him daily, and sometimes more than once in a day, for it was not only at night that he manifested himself. Once I found him standing behind my chair as I lunched, once I saw him in the passage leading to my bedroom, but most often it was in the drawing-room that he appeared. And by degrees, instead of getting in any way used to the apparition, I found myself regarding it with growing horror.

It never spoke, it never seemed conscious of my presence, still less was it hostile to me; it just appeared and disappeared, silent and unaware. But daily, I suppose, it got into some closer spiritual contact with me, and the sense of something desperate and tragic in solution, so to speak, in the air of my house, but able to become solid and manifest, grew in terror to me. Now, too, I need not look up to see if it was there: I knew that the thing had taken form before my eyes saw it.

Once I woke at night with the knowledge that it was in my room, and, turning in bed, I saw it against the window, where the light of dawn was beginning to show grey.

And yet, mingled with the growing horror, was a fascinated interest. There was, I felt, something more to come, some further manifestation which might explain this semblance of humanity which haunted the place, invisible, apparently, to all eyes but mine and unfelt by other consciousnesses. It came. . . .

I was sitting at my piano one night, at the end of the long room furthest from the door, when the electric circuit fused. I was left in complete darkness, and at once I knew that the apparition, though unseen in the pitch black, was present. Then fear, like the tentacles of some monster, began to fasten on me, and in the darkness I groped my way towards the door. While I was still not half-way there I saw close in front of me, and at about the level of my eyes, a flash as of some small firearm, and next moment I stumbled across something that lay in the empty centre of the room.

I fell full length, and below me was a shape that stirred and quivered, and against my outstretched hand I felt something warm and wet that spurted against it. And then sheer panic, insensate and uncontrollable, seized me, that raised the hair on my head and took from my knees the power of holding me up, and I crawled to the door, wiping off from my hand against the carpet that which I knew was on it. Before I reached the door it opened, a light shone in, and there was my servant, candle in hand.

'The light fused, sir,' he said, 'and I came to see. . . . Have you hurt yourself?'

By the light of his candle I saw that the room was empty. There was nothing that could have tripped me, and on my hand was no trace of anything warm or wet.

A month or two afterwards I met Mrs Aylwin again.

'I succeeded you in your tenancy of No. 12,' I said.

Once more I saw that look of terror cross her face.

'I hope you like the house,' she said.

'I adored the house,' I said, 'but not the tenant of it. I was only there a month—about as long as you, I think. . . . Who was it?'

'The butler of the previous tenant, I believe,' she said. 'He committed suicide.'

'In the long room,' I added. 'He shot himself.'

THE TOP LANDING

I HAD bought that afternoon from a twopenny bookstall a shabby
little volume entitled *Commerce with the Dead*, and I was glanc-
ing through it with growing derision when Geoffrey Halton looked
in. As I considered him to be a dabbler in the superstitious, I read
out to him with contempt the sentence I had got to when he came
in. It was as follows:

The possibility of communication with departed spirits is no longer open
to doubt, but the enquirer will certainly expose himself to grave dangers.
When once he has opened the door, as he may easily do, between the dead
and the living, he cannot be sure what evil and potent intelligences may not
be let loose. They are of terrible power: they may drive the communicator to
madness or suicide——

'Oh, what rot!' said I, closing the book.

Geoffrey was silent a moment.

'It isn't rot,' he said quietly. 'It's deadly and awful truth. I have
seen it happen.'

In answer to my incredulous enquiry he told me the following
story:

'My cousin, Phil Halton, and I,' said he, 'found ourselves a couple
of years ago reluctantly facing the conclusion that, after two idle and
expensive months in London, we must devote August and
September to industry and economy. He had to compose the
incidental music to Shakespeare's *Tempest*, which was to be
produced in November, and I to write a novel for serial publication,
of which at present I had only conceived the outline, though in
detail. But just as we were settling down for August in London Phil,

during a weekend in the country, came across a small Queen Anne house in the little town of Adelsham, which was to be let furnished for the ridiculous sum of five guineas a week, including the ministrations of the cook-housekeeper who was in charge. Naturally, he suspected imperfect drains or leaking roofs, but the agent assured him, as proved to be the case, that all was in perfect repair. So within three days we were established there; I had brought down from London my manservant and a house-maid, and our first dinner in the house convinced us of the abilities of Mrs Ayton. She was a shrivelled little old woman, quick and nimble and silent, and as for the house, if Providence had specially planned an ideal retreat for Phil and me he could not have bettered it. It stood on the top of the hill above the Romney Marsh in a cobbled, sequestered street. There was a sitting-room for each of us, two good bedrooms on the first floor, and above four smaller rooms for servants. One of these was locked, and contained, so Mrs Ayton told us, the more private possessions of the owner, who was spending the summer abroad. To the garden, framed in tall walls of mellow brick, there was the access of French windows from the dining-room and the downstairs sitting-room, in which was a piano at Phil's disposal. All was exactly perfect.

'We plunged at once into our belated labours, and for the first week or so I can tell you of no unusual or sinister happening except that both of us, and that increasingly, had the sense that we were being watched. It seemed a fantastic idea, and we used to laugh at each other and ourselves, but a dozen times a day one or other of us would find ourselves suddenly looking round with the notion that there was some presence other than our own in the room. All the time, too, I was cudgelling my memory to recollect what story I had once read about some dreadful occurrence at Adelsham, but at present it was stored in some sealed cell in my brain, and I could not get access to it.

'We were sitting one sultry evening after dinner out in the garden, and dusk was deepening into the night. We had been talking about this odd fancy that something was watching us, when Phil pointed to the house.

' "Look at the light in that window," he said. "Isn't that the closed room?"

'There, sure enough, on the top storey, the windows of which appeared above the low brick balustrade, was a light coming from the locked room. There was also someone in the room for across the illuminated square appeared the black silhouette of a head and shoulders.

' "Probably Mrs Ayton," said I, getting up. "Come in and play picquet."

'When I looked again the figure was gone. But my heart missed a beat, for I clearly saw that it was not a figure of a woman, but of a man. I felt, too, that it was from there that we were being watched.

'Next morning a domestic bombshell exploded. My servant Manders, who had been with me for ten years, came to me after breakfast and asked if he and my housemaid might go back to London. It was reasonable to ask for an explanation.

' "We can neither of us stand it any longer," he said. "I'm very sorry, sir, but I can't stop here, and it's the same with Edith."

' "And if I refuse?" I asked.

'Manders wiped his forehead with the back of his hand, and I saw that his face was white and his hand trembled.

' "Then we shall both ask to give you notice, sir," he said, "and forfeit a month's wages."

' "What's the matter, Manders?" I asked. "You're frightened. Can't you tell me what it is?"

'He looked round with a scared eye.

' "There's something on the top landing, sir," he said in a whisper. "I don't know what it is, but I can't bear it."

' "Have you seen anything, or heard anything?" I asked.

' "No, sir. But it's there, and it's getting worse."

'He took a step towards me.

' "I wish I could persuade you and Mr Phil to come too," he said. "Don't stop here, sir!"

'It was evident that they were determined not to stop here, and unless I was prepared to lose these admirable people altogether I must let them go. But I tried one more appeal.

' "And what's to happen to us if you and Edith go?" I asked.

' "Mrs Ayton says she can manage everything," he said.

' "You can go this afternoon," I said.

'Manders lingered a moment.

' "I'm very sorry, sir," he said.

'A few days after their departure the sudden spate of work which had descended over Phil and me dried up, and, finding that there was a golf links in the neighbourhood, Phil went out to sample and report on it, leaving me with some overdue letters to write. He returned at lunch-time, rather silent and preoccupied.

' "And the links?" I asked.

' "Oh, they're quite good," he said. "I had half a round with the secretary. He—he told me some strange things."

' "Such as?"

' "Well, he began talking about this house, not knowing we were in it. During this last year it has been let three times, but the tenants never stopped more than a fortnight. And the owner isn't abroad. She's in the house now, and the late owner is in the churchyard. I also found out what you had forgotten about Adelsham. There was a certain Dr Hoart who lived here—actually here, I mean—who conducted curious experiments in raising spirits."

'Then I remembered the rest.

' "And hanged himself," I said.

' "Yes."

' "About the present owner?" I said. "I suppose you mean Mrs Ayton. Did Dr Hoart leave the house to her?"

' "Yes, on the condition that she went on with the experiments they had conducted together."

'That afternoon I sat in the garden, reading over what I had written since I came to Adelsham. So deep was I in it that I leaped from my chair when I heard a voice addressing me, for I had seen no one approach.

' "I came to ask if I might have my evening out after I have served your dinner, sir?" said Mrs Ayton. "I have a bit of business."

' "Certainly," I said. "But are you sure you don't want help? You've got a lot to do, Mrs Ayton."

'She looked at me with a secret sort of smile on her withered old face.

' "No, I can manage," she said. "Two such quiet gentlemen, one with his pretty music and one with his pretty story."

'I made up my mind to spring a surprise on her.

' "You managed for Dr Hoart?" I asked.

'Not a tremor betrayed her.

' "And who might Dr Hoart be?" she asked.

'The evening became dark and overcast, and when we had finished our dinner we went upstairs to the little sitting-room where I worked. The staircase was very dark, for Mrs Ayton had drawn the curtains in the hall, and the only illumination came from the skylight above the top landing where was the locked room. As I fumbled for the electric light I looked up and saw dimly, but distinctly, the figure of a man leaning over the banisters at the top. Next moment the light shone out, and there was no one there. . . . I cannot describe the awful shock this gave me, and it required every ounce of my courage to run upstairs and look into the rooms there. Two, lately vacated by my servants, were empty; the other two, the closed room and Mrs Ayton's room, were locked. But again, more vividly than on that first evening, I had the sense of being watched. I said nothing of this to Phil, and presently we were engaged over picquet. We went to bed early that night, each confessing to a weariness and oppression, and for my part I fell asleep instantly. I woke to find my room lit and Phil standing by me.

' "Something is happening," he said. "Listen!"

'From the landing overhead there was the sound of muffled footsteps and a drone of voices, two of them. Then there came a sudden loud creak from the banisters above and a noise as of rustling leaves. . . . We hurried out on to the lit landing, and there, convulsed and quivering, was suspended a woman's figure. It turned as it oscillated to and fro, and we saw who it was—who, with eyes and tongue protruding, dangled there.

'Before we could get to her came the crowning horror. The rope was pulled up again from above by some agency there, and once more the body descended and jumped on the cord that was round the neck. And then the quivering and convulsions ceased.

'We ran up to untie the rope, but it was too late. The two rooms that had been locked up were open, and there was no one within. In the one there was just the furniture of an ordinary bedroom; the other was absolutely empty.'

SEA MIST

ALL classes in the little town sincerely sympathized with Mr John Verrall in this terrible domestic tragedy which had befallen him, for he had long enjoyed their well-merited respect. For over twenty years there had been no citizen more looked up to, for his integrity, his generosity, and the untiring zeal with which he devoted himself to their interests. He had been born and bred there; his grocery business in the High Street, which he had inherited from his father, was a model of cheapness and excellence; and, like a sensible man, he served behind the counter when his other duties permitted. These were onerous, for he had long been a Councillor of the Borough, then Alderman, and was now Mayor. Caroline, his wife, was a discreet and dignified Mayoress, taking a keen presidential interest in the Girl Guides, in the female inmates of the workhouse of which she was visitor, and in the hospital. She had not the geniality of her husband, but it was seldom that she missed a meeting of the committees of the institutions which were in her province. She was older than him by, perhaps, as much as ten years, but at sixty she was still a woman of energy and physical vigour.

They were both of them ardent naturalists. Not long ago the Mayor had presented his great collection of butterflies and moths, housed in a handsome cabinet, with glazed and cork-lined drawers, to the local museum, but he still pursued his hobby, and occasionally added new specimens to the orderly rows, or replaced dilapidated specimens with better ones. Caroline was a botanist, and often on fine afternoons, when their duties were done, husband and wife set forth together over the reclaimed marshes that stretched southwards from the town to the shingle-banks along the coast on these expeditions. He carried a butterfly-net, and his pockets bulged with nests of glass-topped pill-boxes for the reception of his captures; she had her tin case for specimens to add to her presses of the dried flora of Hampshire. The two were childless, but with their simple and industrious mode of life and their keenly pursued hobbies, which required much mild walking in the open air, they might surely look forward to a harmonious and extended evening of their days.

It was in the late spring of the year that this tragic accident happened. The Mayor and his wife had set out from their detached house below the town, on one of their long walks. Half a mile away, seawards, was a ruined castle, built in the time of Henry VIII to check the incursions of the French. The circular keep was enclosed within an angled line of fortification, and in these outer walls there mounted a flight of stone stairs to the level of the loopholes forty feet above, from which molten lead and other deterrents might be poured on the besiegers. The gallery and the inner wall at the top of these stairs were broken away, so that the last step overhung the void below. This castle was a favourite hunting-ground of Mrs Verrall's, for the crumbling walls and fallen blocks of masonry gave harbourage to many stone-crops and to those kindred plants that find nourishment in the interstices of ruins. After an early cup of tea, the two started in this direction.

Three hours later, as the dusk of the evening was beginning to fall, John Verrall returned alone. He found that his wife was still out, but that was nothing to be surprised at. He told the handsome young woman who was their general servant that, as often happened, they had parted company. Caroline had wanted to prowl about the castle, while he had gone on to search the line of willows and alders that grew beside an adjoining dyke, for caterpillars. Indeed, he was in high good humour, for he had found a couple of caterpillars of the very rare alder moth, and, while waiting for her return, he put these, with a good supply of their food, into one of his breeding-cages. But still she did not come, and, after he had had a bit of cold supper himself, he began to grow uneasy. Night fell, and the moon rose, and now, with a more definite fear that something untoward had happened, he rang up the police station.

No; she had not been seen in the town, and presently a constable arrived, and together they went to the castle, where her husband had last seen her. Possibly, clambering about, she might have sprained an ankle, and was lying there unable to move. Luckily, the night was warm, and she would not have suffered from exposure. The moon was large and full, but it was as well that the constable had brought his lantern, for presently a thick bank of sea-fog formed overhead, obscuring the light. Ten minutes' brisk walking brought them to the castle; they called and shouted, but none answered, and soon their search disclosed her lying all crumpled up just underneath the end of the broken staircase in the wall. Her head must have hit some block of stone on the ground, for the skull was terribly shattered.

An inquest was held, and the manner of her death was easily arrived at. From the position of the body, it was evident that she must have slipped when standing on the top step, forty feet above, and death was instantaneous. Her husband related how he had left her at the castle that afternoon, and in answer to some painful but necessary questions, said that he knew of no trouble on her mind; their married life of over twenty years had been a most happy one. The coroner, after recording the verdict of accidental death, had expressed the deepest sympathy with the widower. He suggested, also, that a barrier ought to be placed across the top end of the staircase in the wall, in order to avoid all possibility of such a lamentable accident occurring again.

John Verrall was wise enough not to permit his bereavement to interfere with his duties. There was no use—indeed, it would have been far worse than useless—in shutting himself up and brooding over his loneliness, and, as soon as the funeral was over, he resumed the full activities of his office. His widowed sister, who lived in the town, came to stay with him for a week, to sort out poor Caroline's possessions, but when she hinted that she would be quite willing to make her permanent home with him, he had no hesitation in refusing her offer, for she was one of those sweet, smiling folk who diffuse depression round them like influenza.

'Very kind of you, Amy,' he said, 'and I'm sure I appreciate your intention. But nobody can be the companion to me that Caroline was. I shall be better alone.'

'But all the housekeeping, the endless little jobs, dear John?' she said. 'You, in your busy life, cannot find the constant supervision——'

'That will be perfectly all right,' said John, firmly. 'Harriet Cox has been here for ten years, ever since she was a girl, and she knows my ways. She requires no supervision.'

So Mrs Reed went back to her own house, with a new dress that had belonged to Caroline, and some under-linen and an amethyst brooch. John found that Harriet Cox was an admirable housekeeper, and made him most comfortable. She worked with far greater alacrity now that she was responsible and unsupervised, her cooking improved, the weekly bills were lower, and the house was resplendent with cleanness and polish.

The spacious strip of garden behind had been Caroline's care; a man came in for an hour or two twice in the week to help with the heavier work of digging and lawnmowing. One evening, strolling here before his supper, John thought he must make some changes. It had been Caroline's fancy to have a border of wild flowers. A trellis twined with honeysuckle stood at the back; there were clumps of tall ox-eye daisies, and red and white valerian. There were campions and loosestrife, and in front, lowlier herbs, harebells, and snapdragon and bugloss; and beyond she had made a rockery of such plants as flourished in the crannies of walls. This was her latest creation, and, no doubt, the contents of her botanical case, gathered in the last hours of her life, would have been added to it. As he paused opposite this pile, John Verrall felt a sudden qualm of distaste for it. Her transplantings had not flourished; they were starved and unhappy specimens, and the conglomeration of rock and slag, out of which they grew, was an eyesore, rather than a decoration. He called to the man who was working in the strip of kitchen-garden beyond.

'I want you to remove that bit of rockery,' he said. 'You might set about it this evening. Just cart the stones away: it won't take you an hour.'

His eye fell on the border of wild flowers. They, too, reminded him of Caroline's walks over the marsh.

'And you had best make an end of that border,' he added. 'It's but a collection of weeds. We'll plant a rose-bed there in the autumn.'

The day had been very hot, and, as often happened in the cool of nightfall, patches of mist began to rise over the surface of the fields. They were quite shallow, for John Verrall, sitting with his pipe in the veranda after his supper, could see the tops of the castle walls stand out black and clear above them against the fading radiance of the sunset. He read his evening paper for a while, then, looking up again, he observed that the mist was drifting thickly over the garden. There was a figure, dimly outlined, bending down by the end of the wild-flower border, just about where Caroline's rockery had stood: the job of removing it must have taken the gardener a good bit longer than he had expected. John rose, for the air was getting chilly, and went indoors. Harriet Cox was just bringing his tray of whisky and water into the sitting-room, and, having locked up the house, she returned for the game of bezique which she and her master often played before bedtime. She had picked it up wonderfully quickly, and it passed the time very pleasantly.

Caroline had had no head for cards at all, and she never would learn. She preferred to sit close to the light, doing her knitting or crochet, and the evenings were wearisome. John was no great hand at reading, and, yawning over his book, he used often to glance up at her, wishing she would take herself off to bed. She had habits that irritated him, and yet he found a secret luxury in observing them. She had a genteel way of running the tip of her tongue along her upper lip, and then slightly opening her mouth as if about to speak. Incessantly she made small clucking noises in her mouth, for her denture did not fit as well as it should, and for the last year her digestion had not been perfect, and there were other little noises. When these occurred, she put up her hand to her lips, and said 'Pardon'.

Certainly the evenings now were more entertaining with these games of bezique with Harriet, in all the bloom of her thirty years, a handsome, buxom woman, full of enjoyment and laughter as she laid down her sequence or her four aces.

'Two hundred and fifty,' she said gleefully, 'and a hundred to say. Well, I am having all the luck tonight.'

They were just coming to the end of their second game, when the sharp trill of an electric bell sounded.

'Must be the front door,' she said, and bustled off to answer it. She had left the door of the sitting-room open, and John heard the drawing of the bolts and the click of the key. Then there was silence, and presently the front door closed again, and she came back.

'That's an odd thing now, Mr Verrall,' she said. 'Not a soul there, and I looked up and down the road, and went to the back door as well. I'll have a peep into my kitchen, and look at the indicator, so we'll know what bell it was that rang.'

She was back again in a moment. 'Not the front door at all,' she said. 'It was the bell of the mistress's room, for the disc was swinging still. Something a bit out of order: it happened like that only yesterday. I'll have it seen to tomorrow. Me to draw a card? Well, I never! Just what I wanted!'

John did not sleep well that night, but he was at last dropping into a doze, when a sharp rap at his door awoke him. His Cairn terrier heard it too, and disliked it, for she jumped out of her basket, barking furiously. He turned on his light, and got up to see what it was: the light shone strongly into the passage at the head of the stairs, but there was no one there. But did something brush, ever so lightly, against the sleeve of his pyjama-jacket? He laughed at himself for letting such a notion even enter his head; it was just a breath of the draught coming up the stairs. But Patsy was uneasy; she would not settle down into her basket again, but whimpered at the door, asking to be let out. She was a spoilt little lady, who usually had her way, and once more he got out of bed, and heard her padding down the stairs to a favourite mat in the hall.

These nocturnal disturbances had shrunk next morning to their due insignificance, and John, recalling them, thought of them as more like some fragment of a dream than waking realities. He walked up into the town to preside at the Borough Bench, and found there was a longish list of cases. Most were of the usual type: motors or motor-cycles being used with expired licences, or being left without proper lights; and two small boys were charged with having broken the barrier lately erected across the top of the staircase in the castle. They were seen clambering about near that fatal edge, and, lying on the ground below, were the broken fragments of the barrier. But there was no real evidence to connect the boys with it, and the Bench dismissed the case. John Verrall, as usual, had been absently scribbling on a sheet of paper on his desk as he listened, and he saw, to his surprise, that he had been making sketches of stairways leading up into the air, and then stopping. Afterwards, he had to see the Town Surveyor over some plans for new houses, and told him that the barrier must be put up again at once. It ought, evidently, to be constructed much more solidly; it should be cemented into the masonry of the walls. A very dangerous place: it would be too hideous if another tragedy occurred there.

He found that his gardener had been busy when he got home. The

stones of the rockery were piled behind the toolshed, and the border
of wild flowers was uprooted. It had been a queer idea of Caroline's
to bring into the garden what belonged to fields and hedges, and he
looked with satisfaction on the empty bed and the demolished
rockery. They had always reminded him of her in some specially
intimate way, and with their removal it was as if some site in his
mind had been cleared for a new and more decorative planting. How
old she had got in this last year, how oppressive and irritating her
presence had become to him. . . .

The air over the fields was trembling in the heat, and the outline
of the castle seemed to waver. Never, since her death, since the day
when he had returned alone with the caterpillars of the alder-moth,
had he been there. But it was time to have done with the past—the
removal of Caroline's garden seemed to assist in that obliteration—
and some day, soon, he would go there again and see that his orders
about a more stalwart barrier at the top of that broken staircase had
been carried out. Perhaps he ought to have gone there before and
faced the associations of the place, for he knew that in his mind
there had now formed a little black pool of terror, in which was
reflected the spot where her crumpled body had lain. That must not
be allowed to spread further, it must be drained off.

There was Patsy, lying in the shade of the tool-shed; sensible little
lady; she got a breath of air there, and was out of the sun. At the
sight of her, the memory of that knock at his door last night, which
woke him up and inspired her to frenzied barking, came back to his
mind with a queer vividness. Patsy had always feared and disliked
Caroline; she would steal out of the room if she entered it; she
would go dinnerless sooner than receive her food from those hands;
if Caroline was planting her rockery when John passed through the
garden with his dog, Patsy gave her a very wide berth.

It was lunch time, and he called: 'Come and get your dinner, Patsy,'
and the dog followed him up the garden, keen for her food. Then an
odd thing happened. As they approached the place where the
rockery had stood, Patsy stopped and began barking betwixt fury and
fright, with her eyes glaring at something there. Then she slunk away
in cover of the privet-hedge, and raced for the house.

Just for a moment, that little black pool of terror that lurked
somewhere deep down in John's consciousness, grew larger and

spread in every direction, for the hot, still sunshine seemed suddenly charged with the very essence of Caroline. But he called common sense to his aid; it was just the sight of Patsy slinking away like that, as he had so often known her to do if Caroline was near, that had conjured up the fanciful illusion that she was there by her demolished rockery and her uprooted bed of wild flowers. All was a figment of his imagination, he said to himself; he was like a child who invents something to appal, and then is scared at it. There must be an end of this; he must revisit the castle without delay. This very afternoon he would go out with his butterfly-net and his nests of glass-topped pill-boxes, and follow, step by step, that last walk he had taken, with Caroline for companion, as far as the castle. Patsy should come with him this time; she had not been with him before, as she would never go walking with Caroline.

They set off through the garden. All memory of the morning's agitation had vanished from Patsy's mind. She found an interesting smell of some sort, where the rockery had stood, and pursued it rapturously into the seeding asparagus bed. She had a swim in one of the dykes, she excavated for a mole without reward, and soon, close in front of them, was the castle. How wise he had been, thought John, to make an effort to take himself in hand like this, for the sight of the place awoke no emotion: it was only interesting to see it again. Just there he and the constable, following the beam of the lantern, had come upon the body; he remembered, with extraordinary distinctness, the gleam of the wet blood by the head. Perpendicularly above it was the undefended edge of the staircase. It was almost laughable to think that for all these weeks he had felt this black, secret terror. Now that idle fancy was dispersed; not a shudder, not a tremor came near him; indeed, there was a certain curious fascination in renewing his memory of the place. With a shrug of his shoulders, he walked round the keep; he netted a specimen of the scarce comma-butterfly, and so came to the entrance of the castle again.

Patsy had left him, and he thought she must have gone out in front of him. But there was no sign of her, and he turned back and whistled to her. For answer there was a series of terrified barkings. They seemed to come from somewhere above him, and he ran to the foot of the staircase and looked up. There she was, near the top of it, and when she saw him she whimpered imploringly. Why did she not come down, he wondered. There lay the stairs, open in front of her; she had but to scamper down them and rejoin him. But she squeezed herself against the wall and tried to slink by, and now John knew that there was something—or was it somebody?—invisible to him, which she could not face. She turned back and crawled up another step. She lay there panting, she whined to him to rescue her. 'Patsy, come along, come along!' he called; but his voice was no louder than a whisper; and now he knew that he could no more go to her than she could come to him. Then came the end: with one despairing howl, she bolted up the few remaining steps and jumped into the void. He ran out, and there lay the little dog, her head dabbled with blood, but beyond the reach of terror.

John Verrall had but one thought in his head, to be gone from the place. He dared not allow himself to think until he was clear of it, yet where would he be safe from the unseen?

'What was it?' he asked himself, as he stepped out on to the sunny levels again, and there was no need to answer that. Yet how could the dead return? All their trafficking with the visible world was ended. They lay quiet beneath the earth, and it was only the disordered fancy of the living that could associate them again with night and day, with sunlight or sea fog. He, John Verrall, had only got to keep a firm hold on himself to render such terrors impotent. He grieved for poor Patsy, but something—she was always nervous and jumpy—had scared her, and she had fled before it, not knowing, till it was too late, that the stair ended in a sheer edge. Any sensible man would agree that it was pure coincidence that she had met her death at that exact spot.

A few hundred yards outside the castle lay the path across the marsh to the sea, and there were boys and girls going to bathe, and couples strolling and sitting in the shade of willows by the dykes. With Patsy's death-cry in his ears, it was comforting to hear the sound of human talk and laughter, and to know that, if he joined one of these normal, cheerful groups and told them, in plain words, the horror that was stirring in the dark of his mind, like some silent-footed creature stalking its prey at night, they would think him insane. He walked home across the sunny fields, intent on his resistance to the invisible power in which he refused to believe. Harriet, surprised to see him back so soon, asked what had become of Patsy, and he told her that she had gone off on her own account, hunting rabbits in the furze-bushes; that often happened, and she always found her way home again. It would only cause him to picture that moment more vividly if he spoke of it, and the silent-footed creature would grow more alert. . . .

But Patsy did not return that night, and dimly, in many dreams, he saw himself fleeing up flights of endless stairs, pursued by something unseen and unfaceable. Next morning, one of the workmen, sent out to the castle to put up the substantial barrier he had ordered, saw by the name on the dog's collar to whom she belonged, and brought the little broken body home.

Then, for the time, that dread invasion recalled its forces. The summer passed, and never once did any qualm of fear come near him. It was even as he had known: a man needed only to refuse admittance into his soul of these terrors, and they would cease to assault him. Week after pleasant week went by, and gradually John Verrall began to allow himself to look back, with a sense of danger triumphantly overcome, on those two or three days when he had been on the brink of panic, figuring the return of the spirit of the woman who slept below the new, handsome monument in the cemetery on the hill. That terror had made phantoms of its own creating. 'Just shows what tricks a man's nerves can play him if he indulges them,' thought John, as he stepped along to church on this October morning. 'And a rare surprise there'll be for those who come to worship today.'

The surprise made its due effect, and a rustle went round the church when the banns of marriage were read, for the first time of asking, between John Verrall, widower, and Harriet Cox, spinster, 'both of this parish'. He was sitting in a pew next to his sister, who had offered to come to live with him after his wife's death, and he could almost feel how she stiffened with astonishment and disapproval. To be sure, it was only six months since Caroline had met with that terrible accident, but where was the good of waiting?

John Verrall looked straight in front of him as his banns were given out. He felt himself lightly touched on the arm, just above the wrist, and turned to the empty seat on his left. Caroline was standing there, and she faced round towards him, and passed the top of her tongue along her upper lip and opened her mouth as if to speak. The vision lasted no longer than the one swift intake of his breath, and was gone.

He felt the cold dew of terror ooze out on his forehead, but the next moment he had got hold of himself again, and called on all the power of his will to resist and defy. He joined in the hymn that followed, he hearkened to the prayers, and soon the congregation trooped out of church to a cheerful voluntary on the organ. Often there was a little pleasant lingering and chatting outside, and today a friend or two shook him by the hand and wished him well, but they were rather abrupt, he thought, as if the surprise had not been altogether welcome. But he cared little for that; there was something else that called and cried to his attention. That apparition was only one more disconcerting trick played him by his own nerves, a sudden twitch jerked from them by the declaration of his approaching marriage. The parson had called on anyone who knew of a just impediment to come forward. . . . Then he was conscious of a curious sensation as of cold or numbness on his arm, above his left wrist, and he pulled up his coat-sleeve. Exactly where he had felt that touch, there was a bruise, already discoloured and definite in shape, as if three fingertips had been pressed there.

The afternoon turned out chilly, with a sea mist beginning to creep in over the marsh, and he thought he would keep to the house and busy himself with the monthly accounts to be submitted next day to the Finance Committee of the Borough. He ran through the charges: there had been some heavy expenditures on the extension of the water-main to supply newly erected houses on the outskirts of the town; there was the relaying of the gas-pipes in one of the streets; there was a column of miscellaneous items, among which was the erection of a strong barrier at the top of the staircase in the castle. . . . He got up. What could be a better way of fighting the terror that had manifested itself in visible form that morning, and was stirring somewhere in the core of consciousness, than to go out once more to the castle, and vanquish it on the spot from which he knew it sprang. It was putting out its tentacles again, and feeling for him, and he must cut them off at the root, or he would never be at peace. The sea mist was still only thin; he would run no risk of losing his way in it if he went now, and calling to Harriet that he was going out for half-an-hour's walk, he set forth.

Ten minutes brought him to the entrance of the castle, and, skirting round the keep, he came to the staircase in the outer wall. Up it he went, his confidence increasing at every step, for not the faintest reverberation of fear now shook him, and he knew that all that had happened there was dead and done with. At the top stood the barrier; the work had been well done, for the wooden bars were close and solidly built into the masonry, and he leant his weight heavily against it for test. He passed down the stairs again with quick, light gait, and out of the castle.

Quite suddenly the mist grew vastly thicker, both on the ground and overhead, swirling in from the sea. But he felt pretty sure that he knew his direction sufficiently well to steer an approximate course across the intervening half-mile to his garden. Perhaps he would have felt a little more comfortable if Patsy had been with him, for often had they been out together when the mist was fully as dense, and she trotted along home as unerringly as if there had been none. Then the shape of trees loomed through the vapours, and he found himself on the edge of a dyke.

It was puzzling: he could not make out which dyke this was. But the best plan now was to follow it, and on he went. Again something loomed in front; not trees, but solid and square, and he was back at the outer wall of the castle. He walked along it to the gate, and took his bearings afresh. The gate, he knew, pointed straight to his house, and off he set for the second time, and was instantly swallowed up in the mist.

It might have been wiser, he thought now, to have followed the dyke, for though it would have added to the length of his walk, it must have brought him back to the road below the town. So he struck off to the right, confident that the dyke was in that direction, and once more the castle wall confronted him. At that, terror began to stir; it was as if some unseen force was quietly guiding him back to the same spot. It was gaining stronger control over him, perhaps in this grey density the source of it was close at hand, and now, with terror growing to panic, as he conjectured inwardly and unmistakably what that force was, he broke into a run, fleeing from it. He caught his foot in some tussock, and fell; he scrambled up again and set off once more, panting for breath and giving little whimpering, dog-like cries of fright. Just so had Patsy sought his aid against the invisible. . . .

Again, after this blind excursion, he was back at the castle wall, and now, like some in-pouring billow, the controlling force swept him through the gateway and round the keep to the foot of the staircase. He guessed to what ultimate fate it was driving him, and, with a last frantic effort, he tried to dash past that dark entrance. But he was as powerless to escape as a straw whirling in the funnel of an eddy of dark water, and now, with expiring strength, he began to mount the stairs. Memories pierced the curtain of terror in which his soul was enwrapped: now he pictured how he himself had followed Caroline up those steps, down which she was to return no more; now he recalled Patsy attempting to squeeze by the invisible presence that drove her, as it was now driving him, on and up. For a moment a ray of hope dawned as he remembered how, not an hour ago, he had tested the strength of the new barrier at the top, and found it firm and stable. Now he was borne against it; he heard it creak; he felt it give under the stress of the unseen.

It splintered, it cracked; and as he turned over, falling through the empty air, he saw, framed in the dark entrance, the face of her whom he had himself thrust over the broken edge.

Sources

'Entomology': *Windsor Magazine* (August 1925)

'The Peerage Cure': *Windsor Magazine* (July 1926)

'When Greek Meets Greek': *Windsor Magazine* (December 1926)

'Doggies': *Windsor Magazine* (January 1928)

'The Case of Bertram Porter': *Windsor Magazine* (March 1911)

'Philip's Safety Razor': *Pearson's Magazine* (March 1919). Reprinted in *The Countess of Lowndes Square* (1920)

'The Hapless Bachelors': *Pearson's Magazine* (March 1921)

'Dicky's Pain': *Windsor Magazine* (April 1927)

'The Bridge Fiend': *Lady's Realm* (November 1903)

'The Drawing-Room Bureau': *Women at Home* (December 1915)

'Music': *Windsor Magazine* (December 1924)

'Aunts and Pianos': *Windsor Magazine* (August 1926). Reprinted in *The Funny Bone*, edited by Cynthia Asquith (1928)

'The Guardian Angel': *Woman* (April 1928)

'The Queen of the Spa': *Windsor Magazine* (September 1926)

'Desirable Residences': *Good Housekeeping* (February 1929)

'The Puce Silk': *Lady's Realm* (November 1907)

'The Godmother': *Nash's Illustrated Weekly* (6 December 1919)

'A Breath of Scandal': *The Storyteller* (July 1932)

'To Account Rendered': *The Storyteller* (June 1925)

'The Superannuation Department AD 1945': *Windsor Magazine* (January 1906)

'The Satyr's Sandals': *Pan* (20 March 1920)

'The Disappearance of Jacob Conifer': *Windsor Magazine* (October 1927)

'The Return of Dodo': *Lady's Realm* (December 1896)

'Dodo's Progress' (as 'The Progress of Princess Waldeneck'): *Lady's Realm* (May 1897)

'Number 12': *Eve* (10 May 1922)

'The Top Landing': *Eve* (7 June 1922)

'Sea Mist': *Illustrated London News* (20 November 1935)